The Antipodean Express

The Antipodean Express

A Journey by Train From New Zealand to Spain

Gregory Hill

First published 2024

Exisle Publishing Ltd
226 High Street, Dunedin, 9016, New Zealand
PO Box 864, Chatswood, NSW 2057, Australia
www.exislepublishing.com

A CiP record for this book is available from the National Library of Australia.

ISBN 978-1-921497-15-5

Designed by Nick Turzynski www.redinc.co.nz
Typeset in Garamond Premier Pro 12/15
Printed in Hong Kong

This book uses paper sourced under ISO 14001 guidelines from well-managed forests and other controlled sources.

10 9 8 7 6 5 4 3 2 1

For Roswitha

Contents

MAPS

Route of the Antipodean Express

Dramatis personae

GREGORY HILL was born in Auckland. In a land where sport comes before all else, he quickly set about establishing his credentials in other directions. A rugby ball ended up in his hands during a primary school match while he was running the wrong way. He nicely gave it to a shouting young gentleman in the opposing jersey and has never touched a ball since.

The French horn was a unique substitute and came to inflame his passions, rule his life and pay his bills until he retired in 2018. Study took place at the University of Auckland and Victoria University of Wellington where he became an Artist Teacher many years later. He moved into full-time orchestral employment before finishing his studies and was subsequently a member of professional orchestras in Auckland, Sydney, Melbourne and Hobart, and a lecturer at the Tasmanian Conservatorium of Music. Later he joined the New Zealand Symphony Orchestra in Wellington as a Principal Horn Player, where he stayed, intermittently dreaming of other places and open spaces, for 31 years.

Living in a country with only the most basic passenger train service, long-distance train travel was an unattainable background fantasy which briefly burst into flower in that halcyon year and a half, between career end and Covid beginning.

Gregory has two adult sons who share a remarkably brief cameo role in the first chapter of this book, and two grandsons. He lives in Wellington with his violinist wife Anne who is the co-star of the whole enterprise.

ANNE LOESER emigrated to New Zealand in 2000. She studied violin at the Hochschule für Musik und Darstellende Kunst in Frankfurt, and later played in professional orchestras in Bonn and Wiesbaden. She also had a busy professional string quartet life based in Frankfurt and Budapest for several years, with performances throughout Europe and in Chile. It appeared that Anne was destined for a long and successful career in the German classical music industry, until one fateful morning in Bonn. Sitting at the breakfast table, she came across an advertisement hidden in the back of a music magazine for a violin vacancy in the New Zealand Symphony Orchestra. Anne had no idea that there were orchestras or even music in New Zealand. However, New Zealand being the tourist beacon that it is for Germans, she decided to give it a go for a little while. A persistent man in the horn section at the back of the orchestra made sure that she stayed for ever. Since then she has become a focal point of music-making throughout New Zealand.

Coming from a country with the most extensive passenger train service imaginable, long-distance train travel was a background inconvenience of no interest to her at all, and she thought others' antipodean fantasies would remain safely unrealized. However, when the call came, she jumped on board with gusto and insists she has no regrets.

Antipodes

41°18'55"S, 174°47'4"E

41°18'55"N, 5°12'56"W

Introduction

Tap tap tap!

'Okay, let's take it from the top again! Second violins, you're too slow. Try to keep up!' Orchestra rehearsal. We were in the cramped and noisy orchestra pit beneath the stage of the Sydney Opera House, preparing for The Australian Opera's next production of the 1980 season. I was an impressionable young French horn player in the ponderously titled Australian Elizabethan Theatre Trust Sydney Orchestra. While the Maestro whipped the second violins into shape, my mind lowered a curtain on the opera house and pulled up a desolate scene of tundra and snow with a long train snaking through it.

A horn-section colleague had just been regaling me in the coffee break with his madcap plan to travel on the Trans-Siberian Railway across Russia from Khabarovsk to Moscow. A sort of back-door entry to Europe, and rather difficult to achieve at the height of the Cold War and the recent Soviet invasion of Afghanistan. My colleague, a rather eccentric chap with a twirly moustache and a Citroën DS car, was nearing retirement and could afford the time and money to attempt the journey. I was at the very start of my career, could afford neither, and listened in envious awe. The remainder of the day was consumed by visions of birch trees, balalaikas and borscht instead of the reality of crotchets, quavers and clefs. Thank goodness for the wayward second violins.

I moved on from that orchestra, and I didn't keep up with my erstwhile colleague. I don't know if his dream came to fruition, but he had certainly lit a slow-burning candle in me. One day in the distant future, when I no longer have to turn up in the pit or on stage every day, I would find myself on the Trans-Siberian Railway.

•

Twenty years later I met my wife to be, Anne, and I flew with her to Germany to meet her family who live there. The Trans-Siberian back door to Europe floated to the surface again and flowered into a potential train trip all the way from Wellington, where we were now living, to Anne's home town of Ludwigshafen am Rhein.

There was an old primary-school atlas anchored in my memory, haunting me. I remember it had a page with New Zealand superimposed on its upside-down antipodes, Spain. The antipodes is the place at the exact opposite point on the globe to where you are standing now. Wellington corresponded fairly closely with Salamanca. This always intrigued me. The train trip fantasy gradually extended beyond Ludwigshafen to Wellington's approximate antipodes in Salamanca.

Then the arrival of Google Earth led me to work out the exact coordinates of our house. From there it was a no-brainer to find the same latitude north instead of south, and the same longitude plus or minus 180 degrees, and there was a spot in Spain, absolutely directly beneath the floorboards of our house. It was somewhere next to a motorway in the middle of nowhere, halfway between Salamanca and Valladolid, in the region of Castilla y León. Zooming in a bit revealed a village nearby. Alaejos. The antipodes of our living room was in a wheatfield a couple of hundred metres north of Alaejos.

This wheatfield became the destination of our railway journey. It's 13,000 kilometres away straight through the Earth under our feet, but we travelled 38,000 kilometres to get there in 78 days; 31,000 kilometres of that was by train.

We live on an island at the southern end of the Pacific Ocean, so there were compromises to get to our antipodes by train, but remarkably few. The only flights (they were unavoidable) were in and out of Australia. In South East Asia there were some buses and a ship too. We made a number of detours from the route by road, but we always returned to the last train stop to continue the rail journey. From Vietnam onwards, it was rail all the way on the uninterrupted thread of steel from Ga Sài Gòn, Saigon Station, through Lhasa to Salamanca for Alaejos, and, incidentally, on to London Heathrow Airport Terminal 2.

Sunday, 31 March 2019

Wellington to Auckland

Wellington in the dawn on an autumn Sunday. The centre of town is lifeless. The inner-city apartment dwellers haven't stirred yet, the earliest churchgoers are still finishing their breakfast, likewise the weekend sportspeople. The sun creeps out from behind the Ōrongorongo mountains, uncharacteristically bright and searching. Its low elevation illuminates a little knot of isolated activity at one side of the imposing brick 1937 Railway Station: a hundred or so people milling around a train.

Anne and I were there with them, to board our Antipodean Express, departing Platform 9, all stops to Alaejos. As befits a departure for the longest-possible journey, we had a farewell committee of two: my adult sons Rohan and Francis. Today's train, the Northern Explorer, operated by KiwiRail on New Zealand's narrow 1.067-metre gauge, would take us to Auckland, my old hometown: about as far as you can go by train in New Zealand. Eleven hours, 680 kilometres to the north.

We were at the station half an hour early but our carriage was already almost full. We had booked two window seats facing each other over a table. In the adjacent aisle seats, at the same table, were two gutsy (one more so, one less) ladies travelling to Papakura. Gutsy number one suggested (or demanded?) that each couple sit together, them forwards, us backwards. Well . . . okay.

We quietly started moving, backwards, right on time at 7.55. My boys gradually diminished in a welter of smiles and waves, then disappeared, filial duties discharged. I was in a thrill of excitement that the past year of planning had finally come to fruition. Now let's see if it all works . . .

A slow trundle through the marshalling yards, out for a salute to Te

he Northern Explorer at Wellington Station.

Whanganui-a-Tara, Wellington's gloriously shining harbour, and into the long dark tunnel through to Tawa. The elevated line along the Paekākāriki coast followed, and there were the high forest-covered flanks of Kapiti Island to the west lighting up for the day, with D'Urville Island a more distant presence on the horizon. I spent the first hour or so listening to the ladies bantering about gardening, cooking, families and husbands; quite entertaining, really. I soon started to realize that I was already engrossed in this journey, watching familiar places roll by (albeit backwards). I couldn't stop looking out the window and concentrating on where we were. No time to read or fall asleep. If it carries on like this, it's going to be exhausting! This really is travelling for the sake of the journey rather than a destination. Or a concert.

We rolled through Mangaweka, a place of magnificent long vistas, chasms and cliffs, and up onto the Volcanic Plateau at miserable Waiouru: New Zealand's highest railway station, cowering under the white refrigerator of 2800-metre Mt Ruapehu. This was where multitudes of pimply conscripts used to noisily disembark from the train in the middle of the night back in the 1960s and 70s, eager to be turned into men, wear khaki, and learn how to shoot communists. A few kilometres further on, we crossed the Whangaehu River on the Tangiwai Bridge. The bridge was washed out by a lahar from Mt Ruapehu on Christmas Eve in 1953, killing 151 people on the forerunner of this very train. Nobody remembers it now; Tangiwai passed unremarked in our carriage.

The top of Mt Ruapehu was hidden in cloud, but it was impressive enough. Lunch was a salmon salad in a cardboard box, eaten while the train tottered precariously on lofty viaducts slung over gaping ravines. Scores of Danish girls left the train at National Park in a flutter of fair hair and technicolour Goretex, so we said goodbye to the Papakura ladies and helped ourselves to two forward-facing seats in a now almost empty carriage. After National Park the North Island Main Trunk Line spectacularly descends the Raurimu Spiral, a rather compact intestinal route down off the edge of the Volcanic Plateau, doubling back and underneath itself through tunnels and extremely tight hairpin bends. A drop of 140 metres over seven slow kilometres of track.

We passed through 'Taumarunui on the Main Trunk Line' without even stopping, which seemed like sacrilege to someone who first did this trip in 1971. Back then, the midnight stop at Taumarunui, with a scrum for tea and fruit cake in bullet-proof crockery, was a rite of passage. Taumarunui reminded me of the one New Zealand Symphony Orchestra concert I did there in my career, Schubert 5 and others, about 30 years previously. Of

New Zealand North Island
Whangārei
PACIFIC OCEAN
AUCKLAND
To Australia
Papakura
Ngāruawāhia
Hamilton
TASMAN SEA
Rotorua
Te Kūiti
Gisborne
Taumarunui
New Plymouth
Raurimu
Mt Ruapehu
National Park
Napier
Waiouru
Hastings
Whanganui
Mangaweka
Palmerston North
D'Urville Island
Kapiti Island
Nelson
WELLINGTON
Rail
Air

course, there was an ex-NZSO staffer whose parents owned the Taumarunui dairy. Not everyone can say they know someone from Taumarunui...

After Te Kūiti it all became more civilized and north of Hamilton at Ngāruawāhia we crossed the Mighty Waikato River, New Zealand's longest. We followed its broad lower reaches downstream past sacred Taupiri Mountain. Eventually the river turned left towards Port Waikato and the Tasman Sea, but we carried on straight ahead to Auckland. The Waikato was the most practical route for the railway surveyors to follow in the 1870s, south from Auckland into fiercely independent Māori-held territory in the mountainous interior.

Approaching Auckland, my adolescent stamping ground of Glen Innes, and the Montana Winery site where I used to play concerts with the New Zealand Chamber Orchestra, were a blast. Crossing Hobson Bay past the boat sheds to the Parnell Baths and Judges Bay, with the sleeping volcanoes of Rangitoto and North Head on the other side, was spectacular. This is where my roots are. I can remember seeing steam locomotives on this very line, from the Parnell Baths in about 1960.

Soon we pulled into the remains of Auckland Station, where the station building had been repurposed as student accommodation and most of the platforms dug up. Off the Northern Explorer ten minutes early into light but warm drizzle. We dragged our packs up to the Crowne Plaza Hotel in Hobson Street. After dumping the packs, we were immediately off to dinner at Depot restaurant, sitting outside at the bar, the scene of so many post-concert wind-down dinners in the recent past. Odd to be doing it without a concert first. All these work-related things: the hotel, the restaurant, being in Auckland; I felt a bit out of place now that I didn't have to work for it. In fact, I found it hard to believe that this used to be my life until only nine months ago.

Monday, 1 April

Auckland to Sydney

We woke up in the pre-eight o'clock gloom, to a miserable rainy day. New Zealand's largest city sits on a dormant volcanic isthmus where the North Island is at its narrowest, less than 2 kilometres from east coast to west. The place is surrounded and infiltrated by the sea, bringing rain on about 100 days per year, and today was one of them. We grabbed an umbrella and were soon out on the street to Ugly Bagels, a breakfast favourite from orchestra touring days. It was cold enough that they even had the fire going. We actually ran into the ex-NZSO staffer I mentioned, from Taumarunui. Well, that was a surprise. We knew her in Wellington and hadn't seen her in about three years. Now here she was talking to us in the rain on Wellesley Street in Auckland. There was catching up to be done, and this meant that we were a bit late getting to a SkyBus to Auckland Airport, but after a minor panic, we made it. In that rain and cloud, it certainly wasn't a fond farewell to Auckland. We checked in easily at the airport, then immediately got lost trying to find the Air New Zealand Lounge and our gate. If we couldn't manage this, how were we going to get ourselves across the world?

The three-and-a-half-hour flight to Sydney, 2200 kilometres, whipped by in a comfortable blur. We put the clocks back two hours then descended through murk into a surprisingly cool Sydney, 18 degrees. We would be staying with friends in Sydney until the day after tomorrow, when the next train to Adelaide departs.

Through an easy Australian immigration experience and there was the great friend to several generations of Anne's family, Anita, with her husband Nammi, to meet us. We set out on the long drive to their home north-west of Parramatta at The Ponds, over the harbour bridge with the Opera House down to the right, through Chatswood, Crows Nest and further; about an hour. This was all slightly familiar to me from when I lived here 40 years

Arrival in Sydney. Harbour Bridge and Opera House in the distance.

ago. Certainly, the Opera House was where I went to work every day, and Chatswood was the rehearsal home of the Sydney Symphony Orchestra when I worked with them.

Arriving at Anita and Nammi's house, we were welcomed and tucked into the comforting embrace of the Australian branch of a sprawling, world-wide Indian dynasty. After all that sitting, we used up the last of the daylight with a gentle stroll to the pond of The Ponds, and there were even some familiar blue and black pūkeko, native to both New Zealand and Australia, strutting about on the grass. Swamp hens, as the Australians imaginatively call them. Poor ludicrous fowl, in the words of the poet Ian Wedde. Back to the house and a home-cooked slap-up Indian feast miraculously appeared before us. The evening dwindled as I marvelled that we were finally out of New Zealand, and the journey to our antipodes had begun.

Tuesday, 2 April

Sydney and the Blue Mountains

An early start for a trip to the Blue Mountains with Anita and Nammi. The Blue Mountains are a massive 1000-metre-high forested sandstone barrier, 50 kilometres to the west of Sydney. They're part of the Great Dividing Range, the backbone of eastern Australia which stretches from Victoria to Queensland. They're not really mountains at all; rather the opposite: a high plateau scoured with a multitude of 700-metre-deep gorges. Once you're there, a journey through the Blue Mountains begins with a descent rather than an ascent.

The Gundungurra and Darug peoples have been living here for tens of thousands of years. The Gundungurra say that back in the Dreamtime, two half-fish, half-reptile monsters fought an epic battle which slashed the gorges and valleys into the landscape.

More recently the Blue Mountains formed a perfect perimeter barrier to contain the British convicts who began arriving at Botany Bay in 1788. The barrier was impenetrable for the first eleven years, although the convicts' misery gave them plenty of temptation to risk a run for it. The belief was that China was just on the other side . . .

I had thought The Ponds was in the vicinity, but it turned out to be a very long way. Nammi, being a bloke, wanted to rely on the GPS navigator, but Anita was sure she knew a better way so that hurdle had to be cleared first. We eventually made it to Wentworth Falls; there was not much water but a stunningly rugged and gigantic chasm. And the rock formations of The Three Sisters, which do look fantastic in the changing mist. It really was a thrill to see this totally un-New Zealand landscape again, after an absence of half a lifetime. We went down the Blue Mountains Scenic Railway, a mining-infrastructure-leftover from the nineteenth and twentieth centuries. The railway is just under half a kilometre long, but in that distance it drops 178 metres, its maximum gradient of 1 in 0.82 making it the world's steepest railway incline and a bit scary. Our downward trip was badly timed. We shared it with a whole school of hysterically screaming, smelly kids. Once at the bottom, though, we found magnificent repose on the boardwalk through the dark, still cathedral of the Aussie rainforest.

Back up to the top and we picnicked in a park nearby, among the wonderfully fragrant pale gum trees and lots of noisy white cockatoos. Total Australian immersion. After a look at the Blue Mountains Botanic Garden at Mt Tomah, 1000 metres high and freezing cold, we descended to a more habitable altitude, back to The Ponds and dinner at the local pub. The barman insisted that I taste Something Very Special: Panhead Pale Ale from exotic Upper Hutt near Wellington in New Zealand, which he thought was terrific. It is. He was astounded to hear that we can buy it at the supermarket at home.

So cheers to our first day out of New Zealand; it had been a great success. Having lived in Sydney for a year and a half, a long time ago, I had no need to revisit the central city, and the long-awaited trip to the Blue Mountains absolutely ticked all the boxes.

Blue Mountains, New South Wales.

Wednesday, 3 April

Sydney to Bogan Gate

The once-a-week Indian Pacific train to Perth via Adelaide is huge! It fills two platforms, and they shunt it together on departure. Three-quarters of a kilometre long. A rail link from Sydney to Perth via Melbourne was established in 1917, but the journey originally involved at least five changes of train. This was because pre-1970, the railway was cursed with different gauges (rail-widths) in different states, a long-hung-over legacy of Australia's pre-Federation colonial past. One train could not travel right across the country until the standard gauge of 1.435 metres was established from coast to coast, and the Indian Pacific was inaugurated only in 1970. Even then our destination of Adelaide was not on the route; it was connected to the Trans-Australian Railway with standard-gauge rails as late as 1982. The Indian Pacific service developed into a serious passenger transport option, with four departures per week in the early 1980s. Then tumbling airfares saw the train's frequency tumble; it now operates just once a week as a sort of cruise-ship on rails.

We discovered which platform our half of the train was on and walked some vast distance to our carriage. Into our pokey little cabin with its seat facing backwards, and a letter informing us that there had been a rock fall somewhere on the track and we wouldn't be making the scheduled trip through the Blue Mountains but detouring south and through Goulburn and Cootamundra. Well, I had been really looking forward to going through the Blue Mountains by train, but so be it. At least we got there in the car yesterday.

After a while the train was shunted together and we departed for Adelaide, half an hour late. One thousand nine hundred kilometres, just over 24 hours. We entered a tunnel and stopped in it. For ten minutes. Then we started again with a jerk and very slowly wended our way south-west through suburbs that became increasingly unfamiliar. Once we hit the countryside, enthusiasm levels increased considerably as we bimbled through lovely rural Aussie scenery, sitting with a prime view in the comfortable lounge car. There was an extraordinary selection of retirees on board. Every kind of shape you can imagine. Welcome to retirement! This is my world now! Poor Anne is, I

think, the youngest person on the train, apart from the staff.

Our unexpected route took us on the New South Wales Main South Line, through Mittagong where I played with the New Zealand Chamber Orchestra in the 1990s, and on through Moss Vale and Goulburn. Here we

Lounging in the lounge car on the Indian Pacific.

passed the junction for the branch line to Canberra, the Federal Capital, 90 kilometres away to the south-west. Sometime after Goulburn it was dinner time: crispy little poussins and a perfect fillet steak. After dark we skirted Cootamundra, famous as the birthplace of both the eponymous and endemic wattle tree and the legendary cricketer Don Bradman. Both Australian icons. Of course, the Aussies love a sporting hero and in 1997 the Prime Minister anointed Bradman The Greatest Living Australian. Most Australians are still familiar with the 1930s' hit 'Our Don Bradman'.

Cootamundra also marked the end of our journey on the Main South Line. We turned right for Stockinbingal and headed north on the very bumpy freight-only Forbes Line through Berendebba and Daroobalgie to connect with our original route at Parkes on the Broken Hill Line. Anne and I returned to our cabin, where the beds had been made up: Anne above, me below. What had seemed pokey was now just cosy.

We climbed into the comfortable beds, put the clocks back half an hour for South Australia, and I fell asleep quickly. I was woken around midnight, somewhere in the vicinity of Bogan Gate, by incredibly violent shaking and a full bladder. We were already restored to the Broken Hill Line, heading west. I'd missed Parkes, home of the renowned Parkes Observatory and its 64-metre pivoting radio telescope. The telescope was the co-star, along with Sam Neill, of the 2000 movie *The Dish*. Still, at least I was there for presumably appropriately named Bogan Gate.

Thursday, 4 April

Broken Hill to Adelaide

A steward woke us up with a cup of coffee as the sun came up around 7 am, then the train pulled into Broken Hill for an hour. During the bumpy night we seemed to have made up for most of the time lost on our lengthy southerly diversion, but an assortment of things to do in Broken Hill were cancelled. Still there was time for a quick look around the closest part of the small mining town: buildings of one and two storeys, with deep Aussie verandahs, quiet and still and empty in the dawn apart from the occasional passing truck. The quintessential Australian outback town. The place seemed rather dominated by our gigantic silver snake of a train, gleaming in the low red sun and overflowing at either end of the station platform.

We had a quick look at the Musicians' Club across the road (it was locked), and a recce of Chloride Street, then it was time for breakfast on board, mostly while still pulled up at the platform. Probably the best breakfast available in Broken Hill: poached eggs, potato cake and salmon. Eating in the dining car is a comfortable but cramped affair with not too much room to swing a knife and fork.

We departed Broken Hill, finishing breakfast which we followed with a coffee in the lounge car. Then, a shower in our own ensuite which worked remarkably well if you kept a tight rein on flailing elbows, knees and head. Back in 1951 the Australians introduced the first-ever hot showers on a train, on the Trans-Australia Express between Port Pirie (north of Adelaide) and Perth.

As the day warmed up, South Australia proved to be quite red and Martian. Real classic outback, with a few kangaroos on duty. We cruised through Peterborough, between the Flinders Ranges and the Mount Lofty Range, once an important and complex crossroads in South Australia's triple-gauge (!) railway past. The train doesn't even stop there now. Things started to become a bit more agricultural but predictably dry and scrubby. Monotonous

South Australian outback.

but mesmerizing. The spell was eventually broken by the announcement of lunch and soon we hit the northern outskirts of Adelaide. We arrived on time just after three o'clock at Adelaide Parklands Terminal, on a platform long enough to take the whole train.

A $15 taxi ride got us to our Airbnb cottage near the corner of Carrington and Hutt streets on the edge of the CBD. A pleasant little old one-storey brick conjoined place with just enough room for us. This would be home until Sunday, when the next train to Darwin departs. We wondered about a trip to Kangaroo Island off the coast tomorrow. It's a very long day-trip, and now after a bit of research it appeared that we'd left it too late to organize it. So we had a quick exploration of the locality, then dinner just around the corner in one of several local Italians. This was a fairly sociable part of town; the pub next door did its best to keep us awake, but after a night on the train we were immune to anything it could throw at us.

Friday, 5 April

Adelaide and Hahndorf

A very slow lazy morning in a real unmoving bed. We decided to take a trip to Hahndorf, some distance out of town. Adelaide was having a normal hot day in the low 30s, though it cooled a bit as our bus climbed up into the Adelaide Hills.

I'd been here before; somewhere near here is the house where I first heard on the radio about the Air New Zealand Mt Erebus disaster in Antarctica back in 1979. Memories of quite extensive work trips to Adelaide in that year with The Australian Opera, and later with the Tasmanian Symphony Orchestra, came back to life. Hahndorf is an apparently typical Aussie rural one-street town but with a big pseudo-German influence, from its original settlers.

We hit the pub for lunch, sharing a large hunk of pork Kassler. The Chinese couple next to us went for the $99 German Mixed Grill which completely defeated them. Both the solid bloke and the sparrow-like girl seemed to be getting stuck in, but by the time they stopped eating, there wasn't a dent in the mountain of food. And the girl's face as she tried the tasting plank of five or six German beers was a study in puckered sourness.

Meanwhile, on the wall opposite me, I read about the visit to Hahndorf of the Duke of Edinburgh, Victoria and Albert's second son, in the late 1860s. It seems he got a huge German reception in the street under the bunting outside the hotel, an oration in German and a specially written song in German: 'O Euer Prinz von Deutschem Blut!' — Oh You Prince of German Blood! The Prince of German Blood, the Duke of Edinburgh, was actually Prince Alfred, Captain of the frigate HMS *Galatea*, presumably on the same voyage that took him to Wellington in 1869, where Mt Alfred, an insignificant bump on the shoulder of Mt Victoria, was appropriately named after him. It was the first visit to New Zealand by a member of the royal family.

Now, back in Hahndorf, the sun and the horizon were being blocked out

Hahndorf. Not Detmold.

by an immense brown cloud. I couldn't tell if it was smoke or dust, but nobody else seemed concerned about it. We slipped into a café for some shelter and a Bienenstich (bee-sting cake), along with an Aussie-German coffee which was depressing proof that two negatives don't make a positive. Anne rehatched the Kangaroo Island plan. So it was over the road to the I-site and, yes, we booked a trip to Kangaroo Island for the next day, leaving home about 6 am.

Feeling very satisfied with ourselves, we caught a bus home to our lovely cottage, then walked in the dark to the renowned Central Market, about 45 minutes away. What a selection of fruit, veges, meat and seafood! There were fabulous apples and peaches to try, and samphire, a vegetable I'd only ever seen at Borough Market in London. It's a pity we were not cooking. There was more local charcuterie than you could shake a stick at. And the fish! Whiting, barramundi, mullet and massive local prawns. I ate six enormous oysters then we shared a mediocre paella. It looked great in its big pan, but most of the good bits didn't end up on our plate. We bought fruit and yoghurt for tomorrow's breakfast and walked home.

Saturday, 6 April

Adelaide and Kangaroo Island

We were at the Adelaide Bus Station by 6.15 am. The place was deserted except for a few others going to Kangaroo Island. It wasn't difficult to find our bus; it was the only one running. A two-hour trip, mostly on motorways with nothing special to see, took us to Cape Jervis at the end of the Fleurieu Peninsula south-west of Adelaide. From there it was a pleasant enough 45-minute ferry ride across the Backstairs Passage to Penneshaw on Kangaroo Island; the sea air was a bit cool, so we stayed inside.

Penneshaw is a tiny settlement and port on the north-east coast of the island, whose main purpose seems to be servicing the regular ferries to and from the mainland. There's some holiday accommodation for South Australians who want to escape the merciless heat of the Adelaide summer, and great views over the Backstairs Passage, back to the Fleurieu Peninsula. Perhaps the most noteworthy residents are the small and imperilled colony of little blue penguins, now numbering only fifteen breeding pairs, down from the 'thousands' that the explorer Matthew Flinders reported in the vicinity in 1802. Despite the struggles of the penguins, the rest of 150-kilometre-long Kangaroo Island is something of a zoological cornucopia: Australian sea lions, New Zealand fur seals, tammar wallabies, koalas and, though I didn't see any, presumably some kangaroos.

At Penneshaw we picked up a little rental car for the day, with maps and brochures, and hit the road. The first stop was the small-but-spread-out beachfront settlement of American River. It's famous for its oysters, but the oyster place was closed. A bit further around the coast Kingscote is well known for its marron, a type of small crayfish. Of course, the well-signposted marron shop in the petrol station had no marron, but the shop was for sale. We carried on westwards towards Flinders Chase National Park, through scrubby dry bush country. Considering we were on an island, there were very few sea views here.

Kangaroo Island.

We eventually made it to Hanson Bay Wildlife Sanctuary where things finally lived up to their promise. Big gum trees with lots of koalas sitting around in them. They were quite fascinating to watch for a while, with even the odd spot of movement if you were patient. Koalas are not native to Kangaroo Island; just eighteen of them were introduced in the 1920s. When we were there the koala population had exploded to about 48,000, a direct mirror image of the poor little blue penguins' plight. A bit further along, on the south-west corner of the island, we got to the Remarkable Rocks which were actually Remarkable: extraordinary eroded and hollowed-out shapes on a large scale. I'd been here before, in 1998, and that time it was unbelievably hot with currawongs sheltering in shade with their mouths hanging open. Now it was refreshingly cool and windy, in fact too cool to hang around for too long.

With time to spare, we set out on the two-hour journey back to Penneshaw. Onto the ferry back to Cape Jervis and into the bus in the dark for the long haul to Adelaide. We finally got home about 10.30 pm, barely awake.

Sunday, 7 April

Adelaide to Mabel Creek

We woke up late but made it earlier by putting the clocks back an hour; it was the end of daylight saving. That meant Adelaide's mild winter was on its way, but we were changing our approximately westerly direction since Wellington to due north to Darwin and the tropics. With three days of inactive train travel ahead of us, we walked the four flat kilometres back to the Parklands Rail Terminal, trundling our wheeled packs behind us. The massive silver train with its two bright red locomotives, just shy of a kilometre long, stood at the platform. Inside the terminal building was a huge crowd of mostly retirees sitting among suitcases and guzzling on free bubbly. There was quite a sense of expectation building, for our journey on the fabled Ghan. Over 53 hours to travel 2980 kilometres.

Before 1929, the only way to get from Port Augusta, north of Adelaide, to Alice Springs was on the back of a camel, invariably driven by an Afghan. In 1929 the narrow 1.067-metre-gauge Central Australia Railway between Port Augusta and Alice Springs was completed, and a train called the Afghan Express inaugurated, tracing the route of the telegraph line to Alice Springs. The Afghan Express was too much of a mouthful for the laconic Aussie tongue, and it soon became The Ghan (you can say it without moving your lips) and has stayed that way ever since.

The Central Australian Railway was notoriously unreliable, mainly due to flooding. In 1980 it was closed and a new 1.435-metre standard-gauge route from Tarcoola further west on the Trans-Australian Railway, to Alice Springs, was opened. I remember reading about it in the newspaper, in bars' rest during opera rehearsals in Sydney. In 1982 the standard gauge was extended south to Adelaide. Then work began on a brand-new railway covering the 1420 kilometres from Alice Springs to Darwin, through the Tanami Desert, in 2001. It was a monstrous project, coping with 50-degree temperatures and three-month monsoon seasons. The Ghan finally made it from the bottom of the country to the top in 2004.

This was goodbye to Adelaide. On previous visits I had never really warmed to this oasis of oenology and culture on the edge of the desert, but this time round

South Australia from the Ghan.

we had spent all of our time out of town and it was a blast, with hardly a minute to spare. Unfortunately, I didn't get the opportunity to reacquaint myself with Adelaide's great contribution to world cuisine, the Pie Floater: 'Floater thanks, mate!' — a meat and gristle pie adrift in a puddle of grey-brown pea soup.

Meanwhile it was time to board the Ghan, which was pretty much identical to the Indian Pacific, both operated by Great Southern Rail. Decor in the cabins and corridors is mostly Formica panelling. In some carriages it's woodgrain, but ours was a comforting institutional beige. We moved off almost imperceptibly right on time at 12.15. North Adelaide swept by outside, interminable and ugly. We settled into the lounge car for a long afternoon of reading, writing and looking. It turned into a lovely afternoon up past the southern end of the Flinders Ranges. I couldn't get enough of the view; this was legendary land that I had never seen before. This was what I came here for!

Later the sea appeared on our left. A quick look at a map and I saw we were at the head of the Spencer Gulf approaching Port Augusta. Unfamiliar geography. From here we headed north-west for a while on the Trans-Australian Railway towards Perth. The endlessness of it all was spectacular. The occasional spot of habitation amid vast mind-numbing emptiness. This took us to a magical outback sunset as scrubby growth darkened and the sky morphed into glowing orange.

Soon it was time for another meal, and afterwards we sat in the lounge car playing bananagram; a kind of boardless Scrabble, with a bunch of tough Aussie sheilas Anne had befriended. At Tarcoola the Ghan left the Perth line and we turned right, into the interior towards Alice Springs and Darwin. We settled into our bunks for a good bumpy night's sleep. Sometime in the middle of the night we must have passed through Mabel Creek, without knowing it.

Monday, 8 April

Marla and Alice Springs

The Ghan stopped at 6 am for an hour or so at Marla. From the train, Marla is a weather-beaten station platform with a shed on it, and nothing else. It doesn't appear on any reasonably scaled map. Everybody climbed down off the train into the dirt and, depending on which carriage they were in, walked up to 400 metres to the centre of the train where a bonfire had been lit in the fading darkness and the attentive staff were handing out bacon-and-egg sliders and coffee or tea. We were here to watch the sunrise in the outback, and it didn't disappoint. The huge sky turned to bronze; the details of the immense empty landscape were slowly filled in before being washed out again by the oppressive glare as the day warmed up. With no buildings around, you could appreciate the length of the train. It stretched almost as far as you could see from left to right. Eventually, we all clambered aboard again and slowly trundled off, before the day got too hot. The morning passed in a meditative desert-induced trance. Our speed was limited to 40 kph for this 100-kilometre section of the track by the railway's American owner Genesee & Wyoming Inc.

At some stage we crossed the border from South Australia into the Northern Territory. Brunch time came; we finally got our kangaroo, tender and rare. Eventually, the train sped up and we arrived in Alice Springs, where there were various things that we could do for about four hours. We chose a trip to Simpson's Gap in the MacDonnell Ranges, about 20 kilometres away. Our pink-haired guide was excellent and really knew her stuff about the Arrernte, the local Aboriginal people. We started with a walk of a couple of kilometres to the top of a low hill with a brilliant view over miles of country in every direction. Epic Australian escarpments rose up in the distance. Absolutely magical land. The heat was excruciating, the land completely dry. It seemed really important to move gently and not overexert. Back into the air-conditioned bus and on to Simpson's Gap, we walked along a dried-out

The Ghan parked up at Marla.

creek bed into the Gap, which was an oasis of cool and greenery. Shattered orange rocks towered above us and the Gap was full of fresh still water. There was some birdlife, but not much else.

We lingered as long as we could then made our way back to Alice Springs and our waiting comfortable train. I finally made it to the front of the train to get a photo of the locomotives; it took forever to get there and back. Anne was quite relieved when I appeared again. Alice Springs itself didn't get a chance to leave much of an impression, other than that of a flat hot suburb of nowhere. The town was set up around 1872 as a relay station on the new telegraph line that linked Adelaide with Britain. It straddles the Todd River, which, in that uniquely Australian way, doesn't usually have any water in it. The locals have adapted to this with their annual Henley-on-Todd Regatta, a zealously fought series of races for foot-powered bottomless boats of any ridiculous design, on the dry sandy river bed. The regatta had to be cancelled in 1993 when the river uncharacteristically filled up with water, and possibly crocodiles.

We carried on into the dry and darkening outback, dining on delicate crocodile sausages, which some may see as just retribution.

Tuesday, 9 April

Tennant Creek to Katherine and Darwin

Sometime in the middle of the night I was vaguely aware, through curtains of shallow sleep, of light and noise as the train stopped at Tennant Creek — another of those legendary Australian nowhere destinations. I'd like to have taken in a bit more of it, but I didn't really wake up, and by morning it was a vague memory or maybe a dream.

After breakfast we arrived at Katherine to get off and have a look around. We took a tourist excursion to Nitmiluk National Park to go by slow boat through the Katherine Gorge. The gorge is actually a huge system of gorges on the Katherine River as it cuts through the landscape. We were in large flat-bottomed, open boats that held lots of people. The views of those chasms cut into the Australian rock and eucalypt forest were just marvellous. Compared to yesterday's experience, everything was green and wet — and rocky. The 33-degree heat, with 97 per cent humidity, was debilitating.

This was our first experience of Aboriginals in Australia. After hearing so much about them and their culture, it was starting to feel as if they were hidden away. But here the guides and boat-drivers were mostly Aboriginal. Nitmiluk was a fabulous place to experience; it rather blocked out the rest of the day and I have no recollection of the town of Katherine at all.

We were out of the sauna and back on the train for lunch and the last 320 kilometres to Darwin. Meals were eaten at snug tables for four, randomly with a different couple each time. Maybe half Australians, the rest Americans and Canadians, a smattering of New Zealanders and no Germans. No Chinese either, from memory. Around this time the view out the window was dominated by the most extraordinary anthills: tall narrow towers maybe a couple of metres high or more. I'd never seen anything quite like them.

After a while we were swallowed up by the absolutely vast Darwin railway

Katherine Gorge.

yards, journeyed through them for what seemed an age, and pulled up dead on time at 5.30 pm. The railway's main purpose is freight, so it terminates at the port and the station is some distance out of town. Great Southern Rail put on a free bus into the city, with endless stops at all the hotels. We finally reached ours, the Vibe, on the waterfront in the centre, and checked into a big, light, comfortable if rather plain room.

Now it was time to hit the town and find something to eat. Darwin is a somewhat sleepy town, with some noisy action only in the very centre. Now that we'd weaned from the pre-paid bosom of Great Southern Railway, we had to pay for our food! I was really looking forward to something clean and healthy for a change. The rich food on the Ghan had become a rather guilty pleasure, so we found a good Japanese restaurant for a beautiful platter of sashimi. Just what we'd been missing. Then we walked home, not very far, in a perfect temperature with big fat raindrops plopping out of the sky.

Wednesday, 10 April

Darwin

On opening the curtains, warm bright light poured into the room. We had a view across grassy parklands to a section of the harbour, and the sea. There was a search for breakfast; the Darwin waterfront is absolutely dead at 9 am. Then it was back to the Vibe to arrange printing out of newly arrived Chinese train vouchers and instructions. The helpful Chinese girl at reception who did it all was very jealous of the places in China we were going to visit.

We checked out, dumped the luggage and set off for a long hot walk through the centre of town and to the Botanic Gardens and the museum. Walking was okay if you took it slowly and searched for shade. Darwin is functional rather than attractive; it's all flat and most of the buildings have been built since Cyclone Tracy laid waste to the city in 1974. There isn't really any sense of history at all, apart from a couple of nineteenth-century stone walls held up by steel bracing: the remains of the old Town Hall, irreparably wrecked in the cyclone. There's a very small area by the sea that is the site of the first British settlement, then called Palmerston, but with not a single structure standing. All the public buildings are modern, and we saw a surprising number of apartment buildings, but not many people on view.

There were hints of a nasty undercurrent of violence; there was a big police and security presence, and bus drivers sat in a metal cage, with warnings of closed-circuit security TV on the bus windows. Drunk, unhappy Aboriginals were lounging about the public spaces, mostly keeping to themselves. You might have heard about this all your life, but it's confronting when you first see it.

We eventually made it to the Botanic Gardens for a spot of air conditioning and a drink. Lunch might have been a good idea, but we couldn't face it after the excesses of the last few days. The Gardens were a welcome interlude of tropical exotica, then we headed off to the museum via a pleasant seaside walk, looking out over the Beagle Gulf towards the Timor Sea. The museum building was rather uninspiring to look at, and most of the exhibits inside were a bit ho-hum. There wasn't much history on display. Of course, European

Police chase. Darwin.

history here is sparse, and doesn't go far back, but somehow 60,000 years of Aboriginal history and culture got left out, apart from some interesting art. There was a big display about Cyclone Tracy, which I did find arresting — and a monstrous stuffed crocodile.

We caught a bus back to the city through ordered, leafy Australian suburbs. After dinner we collected our luggage and took an Uber to the airport, for a midnight flight to Denpasar, the main city on the Indonesian island of Bali. We couldn't avoid this flight, as there appears to be no shipping service, passenger or freight, between Darwin and Bali. In fact, even direct flights are surprisingly infrequent. Our Uber driver was a very chatty, enthusiastic, friendly mature Indian who thought he had done well for himself in Australia, though he was driving an Uber. We checked in for our flight, no problem. Then we were told that it would be at least an hour late. Jetstar. Of course. So there was nothing for it but to sit around for a few hours in a big airy terminal on uncomfortable hard plastic seats. Quite difficult at the end of a long day when you'd rather be asleep anyway.

This was the end of our ten days and 4800 kilometres in Australia. It had been a very neighbourly, culturally familiar and relaxed introduction to round-the-world travel, despite the seats right now. There was a tightening awareness that things would be very different tomorrow.

Thursday, 11 April

Darwin to Denpasar

The day started right on midnight with the take-off from Darwin, for a two-and-a-quarter-hour flight of 1800 kilometres. It was an uncomfortable one, with no legroom or service. It wasn't even cheap. I slept a little — and got a very sore bottom. The clocks went back one and a half hours, so we landed in Denpasar not too long after leaving Darwin. Denpasar airport seemed huge, and there were massive queues at Indonesian immigration. Fortunately, nobody was interested in us, and we sailed through.

The first concern was getting some Indonesian currency. There were three ATMs in our vicinity, but only one of them was working, with a long slow queue. Still, a bit of patience, and sometime around 2 am we had a couple of million Indonesian rupiah in our hands. I had been warned that the taxi situation at Denpasar airport could be a nightmare, so we had had the good sense to arrange for a driver from our hotel to pick us up at 1 am. Sure enough, as we came out into the public concourse an hour late, looking around bewilderedly for something familiar, there was our man Mr Agu with a big 'Gregory Hill' sign and an even bigger smile. The drive to our hotel, La Taverna Suites, on the beach at Sanur on the east side of Denpasar, took about half an hour: around the outskirts of the city then into the beachside darkness. We were made to feel very welcome, shown to a comfortable room, and fell into bed and sleep immediately around 4.30. We slept solidly till about 8 am when curiosity stirred us and demanded that we explore our new exotic location. No more Australian familiarity. First our room: on the ground floor of a two-storey plastered building. A big tile-floored bedroom with a fourposter bed with mosquito nets, an amazing stone bathroom, and our own private patio-garden sitting area. All set within lovely well-kept tropical gardens. It was all very low-key, clean, comfortable and pleasant. La Taverna is supposedly one of Sanur's original hotels; it's dated and doesn't reach modern standards of antiseptic luxury and I loved it.

We ventured out to reception to register, under an umbrella in the garden. The staff were the perfect combination of politeness, familiarity and friendliness. A woman led us out to breakfast, past the pool, numerous sitting-about areas and onto the beach. Tables with umbrellas on the sand. Breakfast

Sanur Beach, Denpasar.

with feet in the sand. Somehow, I wasn't expecting that. Only one other table was occupied. Breakfast was fresh tropical fruit, good strong south-east Asian coffee, and a rather dull but enjoyable bami goreng with a fried egg on top.

We were facing the sea with a 180-degree view of it; there might have been more to see beyond that, but the horizon disappeared into the haze. Directly behind us was the narrow 4-kilometre-long walking and cycling path that stretches the length of Sanur Beach and behind that, our hotel. The path was a busy thoroughfare, but with no motor vehicles it was peaceful enough; just an endless procession of people going nowhere in particular — tourists, apparently resident expats, and Balinese. We joined the procession. It was fascinating people watching. Lots of Russian women all cramming the smallest quantities of their flesh into virtually non-existent beachwear, Germans and of course Aussies too. The Balinese were lovely and friendly, although around here their lives were geared to satisfying tourists, and there wasn't much indication of how they really lived. There were a few places where the path intersected with roads leading back from the beach: masses of parked scooters, traffic, election billboards and Hindu shrines and statues.

At lunchtime we shared a yummy fish-in-banana-leaf thing. We spent the afternoon walking further, looking in tourist shops but unable to buy much: where would we put it? Real charcoal-grilled satays had to be tried, and a foot massage, and lounging in loungers on the beach. Lombok Island gradually loomed up out of the clearing mist; a dry thunderstorm came in and the tide went out.

We had dinner next to the beach, gado-gado and roast pork ribs, a sort of small-scale replica of babi guling, the famous Balinese spit-roasted suckling pig. I'd been looking for it everywhere to no avail; this would have to do. There would be no chance of getting babi guling anywhere else in Muslim Indonesia. All in all, a relaxing day at Sanur, as intended, despite the tourists everywhere. Everywhere that is, except our lovely La Taverna, which, despite having about 30 rooms, was very quiet.

Friday, 12 April

Denpasar to Banyuwangi

The plan for today was to leave Bali, heading west across the narrow Bali Strait to Java. To catch the ferry to Java, we had to get from Denpasar in the south of Bali to Gilimanuk at the western tip of the island, about 130 kilometres by road. There are no trains in Bali and going by bus would have involved all sorts of messing about with connections — and slow buses. The hotel manager recommended a car and driver for NZ$100 and we went for it. So, having arranged all that the previous evening, we started the day off with another breakfast on the beach. I had nasi goreng this time, the other option. It was pretty uninspired but was included in the room rate. I asked for extra chilli, so they chopped up a dynamite little red mouth grenade and chucked it on top. Bang!

Our car turned up right on time at 10 am and we said goodbye to our wonderful La Taverna. First stop was an ATM to get another couple of million rupiah. As soon as we left the hotel, we saw that the peace and quiet was a complete illusion and we were in major congested traffic chaos. We had to get out the other side of Denpasar; I don't think we passed through the centre, but the crazy built-up city lasted over an hour, before succumbing to the odd spot of bush and rice paddies, and occasional tantalizing inland views. Our middle-aged driver gradually opened up a bit; once he realized we were receptive he spoke about family and Balinese culture, and he was knowledgeable about gamelan when I asked him about that. Gamelan is the traditional music of Bali and Java, often quite hypnotic, played by anything from a small ensemble to a whole orchestra, of gongs and tuned percussion.

Apart from pleasantries and practicalities with the hotel manager, our driver was the only Balinese we really conversed with. He didn't ask me much, but I guess a lack of individual curiosity about his hundreds of hurrying foreign customers is justifiable. We passed a few babi guling joints on the way, now that I had no need for food. That will have to wait for another time. We stopped for petrol and our man asked me for 300,000 rupiah to pay for it, and I thought that was rather odd and hoped it was included in the final deal. The suspicious

Western mind! In fact, he just didn't have that sort of money until I'd paid him.

The next stop was the town of Negara for lunch. The driver's choice was some ghastly Indo-pizza joint; we certainly weren't going to eat there. We were both still full of breakfast nasi goreng, so in fact we just did without lunch. However, for both of us this town was our introduction to the workings of a squat toilet and the water rituals involved instead of flushing. We carried on north-west through a well-forested national park with grey macaque monkeys by the side of the road. All of this, of course, involved standard kamikaze-style driving, being overtaken while you yourself are overtaking someone else; meanwhile a truck is coming the other way, determined to kill us all. It's best not to look out the front, and just concentrate on the trees and monkeys whizzing by the side windows. The closer we got towards Java the more the Muslim influence grew. There were fewer Hindu shrines, and we started seeing mosques and minarets. Certainly, no more babi guling.

Gilimanuk Port, Bali.
Java in the distance.

Suddenly we were in Gilimanuk and pulling up at the ferry terminal. Of course, the driver stuck to the agreed rate and subtracted the petrol money. Then he helped us with the ticket-buying process, which was just as well because there was no English here, either written or spoken. Tickets to Banyuwangi at the eastern end of Java cost 60 NZ cents each for the 45-minute journey. The tickets got us through the modern terminal onto a long walkway with a good look at a real Asian small port with all sorts of down-and-out looking vessels.

The ferry was one of many; they depart every half hour. It was a utilitarian thing with two open-air levels, cars downstairs. We sat outside at the stern, watching Bali recede. Java was already visible, not far away over the bow. The clientele was mostly Indonesian with just a few European backpackers. The clocks went back one hour to Java time, which added to the sense of excitement, expectation and anxiety on this short voyage from one province, and one religion and social structure, to another.

Disembarking at Banyuwangi led us through more long pedestrian thoroughfares to a smiling boy with a 'Gregory Richard Hill' sign, ready to whisk us away for a fifteen-minute drive to Bangsring Breeze Hotel. Gone were the Bali Hindu shrines; mosques and hijabs were everywhere. We left the coast road, drove through an area of fairly apparent poverty, before climbing to the absolute paradise of Bangsring Breeze. After checking in we entered our villa high up on the property among the trees, with a view of patches of sea and Bali hunched in the distance, as the muezzins calling from the minarets below echoed and intertwined with wafting gamelan music. It was a peculiarly Indonesian Muslim soundscape. Simply magic. This place is an oasis of peace. Soon we were sitting beside the pool with a Bintang beer in hand, listening to thunder rolling around the black afternoon sky.

Bangsring Breeze is well out of Banyuwangi; we never saw the city. And there were only two other guests, a French couple. Back to our very comfortable villa: the huge outdoor stone and plaster bathroom had to be seen to be believed. No toilet paper down the loo, though, and the very impressive-looking rain shower was a little pathetic. Dinner was in the in-house restaurant, and we were in bed and asleep by eight o'clock in preparation for a big day tomorrow.

Saturday, 13 April

Banyuwangi and Kawah Ijen

Mt Ijen is a 2800-metre volcano near the eastern end of Java. It's renowned for the blue sulphurous flames emanating from the jumble of rocks inside its massive, and very active, crater (Kawah). Of course, the flames are only visible in the dark so you have to be in the crater before sunrise to see them.

We were awakened at half-past midnight, after four hours' sleep, and were down at the gate at 1 am and into an SUV with a driver and a guide. Down to the coast road, through the darkened streets of Banyuwangi, then a right turn inland towards Ijen. We gradually became aware of a gentle increase in altitude, cruising through dark villages which gave way to forest, then the road became incredibly steep and narrow. There was nothing to see in the dark and there was quite a bit of fitful dozing. Every time you stirred there was the same seemingly vertical wall of road illuminated by the headlights, in front of us, and gloomy ghostly shadows of trees flitting by right up to the edge of the road beside us. After an hour and a half of this we arrived at Pos Paltuding, where the walk to the summit starts.

It seems Indonesians see Kawah Ijen as something of a pilgrimage, and we joined a big throng of people in a tunnel of torchlight in the pitch dark, trudging up the narrow track, which became steeper and steeper. Soon we got to the so-called taxi-drivers, pairs of men who would push-pull you up in a two-wheeled buggy. The huge crowd was certainly not limited to fit, outdoorsy types. For us the climb was a breathless struggle with a few very short breaks for one and a half hours, then we were at the top. Up here we could still see nothing of our surroundings; the sky had cleared completely but there was no moon. We were on the lip of the crater and couldn't really see how far it extended. Where we stood, a steep rocky trail led to the depths of the crater 220 metres below us. There was a big sign saying, even in English, that it was Not Legal to go down there.

There is a group of people who do have legal access to the crater, however, and that's the sulphur-carriers who climb into the crater and collect large

In the crater, Kawah Ijen.

sulphurous rocks, placed into baskets. Two baskets go onto the ends of a shoulder yoke — a normal load is about 70 kilograms — and the poor guys stagger up to the top, battling their way through the hordes of tourists coming down. Having established that we were not allowed into the crater, Rudy the guide suggested we do it anyway, following the masses, and we hired filtering masks to protect us from the sulphur fumes.

Down the narrow cleft we clambered. It was very rough and slippery and overcrowded. Angry, shouting Indonesian men tried to clear the way for the ascending sulphur carriers, and angry, shouting tourist Chinese men tried to push their way forward as much as possible. It was an unpleasant jostle with tempers flaring everywhere. Indonesian tourists, Chinese tourists, the poor sulphur-carriers, a smattering of European backpackers and us. Eventually, we made it to the bottom. The promised blue flames were scattered among the chaos of rocks. Gradually, the sky brightened, the flames dimmed and the massive crater we were in was revealed to us. Great gouts of steam issued from the multicoloured rocky mess, turning into shifting mist. There was a huge crater lake stretching away to a collapsed wall in the distance where its outlet presumably was. Closer at hand, we were surrounded by the sheer 220-metre walls down which we had scrambled. As the sun rose, the sky lit up with a blaze of orange. We stayed for a long time, astounded by all of it.

We reluctantly retreated up the near-vertical crater walls, in a slightly more subdued queue, past the struggling sulphur-carriers. The sulphur-carriers, on attaining the rim, all sit down for a refreshing smoke in the slightly oxygen-deprived sulphurous air. After a break, they shoulder their laden yokes and trudge 3 kilometres down the outside of the mountain to a refinery in the Paltuding valley. The sulphur ends up in rubber, fertilizer, matches and ... white sugar. In return for severe health risks, constant physical danger and back-breaking hard work, the men earn about NZ$20 a day, if they can manage two trips. It's considered quite good money.

The view from the top, in broad daylight now, was extraordinary: our mountain, and other volcanoes of the Ijen plateau around us, rising out of the early morning mist. We headed back down the mountain the way we had come. As we descended there was still a solid stream of people coming up past us to experience the mountain in daylight. Including, at one stage, the French couple who had been the only other people at Bangsring Breeze. I called out 'Oi! Bangsring Breeze!' The woman acknowledged me, but the nice Frenchman just kept walking.

We got back to the car for a very long drive down rough four-wheel-drive tracks to the north-west of Ijen, eventually ending up at the Belawan Hot Springs, a quite comfortably done spa run by a coffee-growing company, off the edge of the Ijen plateau. We seemed to have the whole place just for the two of us, then a gaggle of Muslim women turned up for a girls' day out at the pool. They might have been surprised to see us in their pool, especially Anne in her bikini, but after a few exploratory toe-splashes they just jumped in fully clothed and seemed to be having a grand time. They were very friendly to us and encouraged Anne to take photos of them. Of course, they wanted to know where we came from. Normally 'New Zealand' is a good answer in most countries, because we are too small to have done many bad things to other people, but this was only a month after the massacre at the mosques in Christchurch, and I felt quite uncomfortable about it. However, they said New Zealand is very good country and they all liked Prime Minister Jacinda Ardern. These lovely women were very accepting of us.

We finally made it back home for a quiet Bangsring Breeze afternoon. By now we were the only guests, and it was beautifully restful. Enduring memories of Bangsring Breeze: the endless religious chanting and the gamelan coming up from below; the birdsong, especially the one which gets faster and faster and higher and higher; the shining sea in the distance; and the peace.

Sunday, 14 April

Banyuwangi to Surabaya

Today we were back on the trains after a hiatus of four days that seemed like forever. It was to be a six-and-a-half-hour journey to cover the 300 kilometres to Surabaya on a narrow 1.067-metre gauge.

A Bangsring Breeze driver took us to Banyuwangi Baru Stasiun. Down along the coastal boulevard beside the Bali Strait we went, for a last look at the mountains of Bali squatting under cloud in the distance. We had vouchers for the train, paid for online and printed out at home, not actual tickets. We scanned the vouchers on the machine at the station entrance, with some help from a friendly policeman, and it seemed a miracle to me that our names popped up on the screen. Press 'print', and now we had bright orange tickets for Train 88, the Mutiara Timur, *eksekutif kelas*, which we showed to the nice steel-helmeted soldier with a sub-machinegun who let us into the station. There were lots of soldiers around, but they all seemed quite relaxed.

Banyuwangi Baru is the end of the line, so our train started from here. There were plenty of people waiting; not another white face to be seen. The station was impeccably clean. You could safely eat your breakfast off the polished marble floor if you wanted to. The Mutiara Timur came in at walking pace, escorted by a policeman: a big, white diesel-electric locomotive pulling a long grey train. Our carriage was very comfortable and spotlessly clean. The trains in Java are operated by P.T. Kereta Api: Fire Carriage Company. Our Fire Carriage departed ten seconds early according to my phone. Gently out of the station past backyards, rubbish, level crossings with big queues of scooters and trucks, mosques, rice paddies, corn, chillies, *warungs* (small eateries), forest, mountainous windy bits, bridges, rivers, tunnels, squalor. Spellbinding.

Lunch came around, served from a trolley in the carriage. A tasty fried chicken breast, red and green sambals, tempeh, rice, prawn crackers, lettuce leaf and cucumber. NZ$3. Gradually, the rural and provincial landscape was lost in the outskirts of massive sprawling Surabaya. We arrived at Surabaya Gubeng Stasiun about ten minutes late and took a taxi through crazy traffic to our Hotel Majapahit.

First appearances were encouraging: a massive spread-out colonial place on two levels, lots of courtyards with fountains and statuary, and colonnaded walkways. It was very easy to get lost in, which we did several times. Our ground floor room was huge, with a palatial bathroom. The toilet door had a beautiful big, bevelled-glass window, perhaps impractical but very stylish.

Surabaya, still known as the city of heroes, was the location of fierce fighting in the struggle for independence from the Dutch, and the British, at the end of World War II. In fact, our Majapahit Hotel, then called the Yamato Hotel, was the scene of a confrontation in that conflict, known as The Flag Incident, and the story is graphically told on a plaque beneath the Majapahit's flag tower:

> *It is here that on September 19th 1945 occurred what is known as the Flag Incident. Angry people of Surabaya requested that the Dutch flag on top of the Hotel Yamato was taken down. S. Kasman, a young man from the Youth of the Republic of Indonesia (PRI), Roeslan and Sumarsono managed to mobilize crowds to gather on Jalan Tunjungan. Kusno, staff of the Surabaya District Office, went through the Dutch and Japanese security ring and climbed to the top of the hotel, grabbed the Dutch flag and tore down the blue section to keep the red and white of the Indonesian flag. Mr Ploegman lost his life during the incident as his body was hacked by machetes and takeyari.*

Having read this and aware of the ISIL terrorist attacks on Christian churches in Surabaya less than a year previously, in which 28 people died, we set out to see what we could of the city in the hour or so left to us before dark. We were on a major thoroughfare of tall buildings and chaotic multi-lane traffic but soon found relative tranquillity beside the Kali Mas river which snakes through the city. There were a few scenic bridges, rundown buildings and greenery; it was pretty decayed and grotty as we walked along the narrow riverside laneways with no footpaths but plenty of puddles and potholes, avoiding scooters. It was all quite fascinating, but we started to feel a bit conspicuous and so headed home in the dark.

We had a magnificent dinner in the hotel restaurant. It was a meaty business: rawon, a black beef soup for which Surabaya is famous. It's darkened by black keluak nuts and served with accompaniments similar to the lunch on

Kali Mas River, Surabaya.
City of Heroes.

the train. It was certainly worth travelling to Indonesia for. Meanwhile, the cringe factor was provided by a table of Australians complaining and sending food back and eating fish and chips and burgers.

Afterwards we had a quick look around the closed shops in the hotel foyer, including an Asian-style over-the-top European patisserie. Western music seemed to be the theme, with rows of different guitar cakes and grand piano cakes. No horns or violins, though. The message written in icing on the chocolate lid of a piano cake: 'I will always be there for you, will you be there for me Happy Valentine's Day'. In April.

Monday, 15 April

Surabaya to Borobudur

Back on the rails again for another 300-kilometre trip, to Yogyakarta, six and a half hours, followed by a one-and-a-half-hour drive to Borobudur. A quick trip to last night's patisserie furnished us with pastries for breakfast on the train, then into a taxi, back to Surabaya Gubeng Stasiun, past the rather incongruous Russian submarine beached in an amusement park. By now we'd learnt the Indonesian railway ticketing system and we breezed through it like old hands.

The station was grubbier than Banyuwangi Baru, and our train, number 101, the Ranggajati, to Tugu Yogyakarta, was nowhere to be seen. We discovered that we must board the train standing at the platform, which was yesterday's Mutiara Timur from Banyuwangi, and get out on the other side onto an extremely narrow platform. The Ranggajati was hidden on the other side of that. Apart from platform 1, the whole station seemed to be made up of these finger-width platforms that can only be accessed through parked trains. You just have to make sure you stop your traverse at the right train, and that the train you are traversing doesn't leave while you're still in it. Just like the station, today's train was dirtier than yesterday's but still very comfortable.

There was an interesting little piece of theatre before departure: the entire train's crew was lined up on the platform, military-style, before an officer who barked orders like a drill-sergeant. There was much standing at attention, stamping of feet and the like. There were bowed heads too at one stage, which suggested something religious. Then the parade was dismissed, and everyone went back to being train staff.

We left Surabaya a couple of minutes early, around 9.15. The scenery was broadly similar to yesterday's, without the novelty. Indonesian trains go through endless paddy fields, mesmerizingly beautiful with reflections on the watery surface. Intermittently, there are working parties of colourfully

Surabaya Gubeng Stasiun.

dressed people, and the odd buffalo. There are plenty of small settlements; they invariably present their worst-polluting rubbish tips to the railway. The amount of uncontrolled rubbish in Indonesia is mind-boggling.

The lunch on this train was pretty rubbish too: a chunk of sweet, spicy, tough beef and some fatty bits and pieces. The weather gradually closed in, with plenty of rain and thunder. We knew we were supposed to go past some impressive volcanoes, but they were only implied as vague shadows in the rainy haze. The sky became really dark in the early afternoon. There were long delays at several stations, and we finally got to Tugu Yogyakarta one and a half hours late at four o'clock.

Our man Ariel was there to pick us up, not at all concerned at having had to wait. He drove us for one and a half hours through solid traffic and commerce out of Yogyakarta to the famous Buddhist temple of Borobudur, and our dreamily relaxed accommodation for the night, Rumah Dharma. This is a smallish complex of luxurious huts in a tropical garden setting among the fields. It appears to be family-operated. We arrived right on the onset of a purple sunset and sat outside on our porch sipping fruit juice. Later we strolled over to the dining area in the garden, in intermittent light warm rain, for a superb Javanese dinner with little fried bananas for dessert.

Tuesday, 16 April

Borobudur and Yogyakarta

We were up at 3.45 am, setting out on bicycles through the pitch blackness to the Borobudur Temple only a couple of kilometres away. Entrance, security and ticket buying were in the large complex of the Manohara Hotel. We strolled on manicured paths through dark gardens towards the invisible temple. At the base of the temple a steep narrow stairway took us in several stages to the top, about 35 metres higher. It has no interior as such, as the whole thing was built on the flanks of a 30-metre hill, around 1200 years ago. The top levels, increasingly smaller like a pyramid, are populated by regiments of bell-shaped stupas of various sizes. We stood among the stupas with a large crowd of tourists from all over the world. The sky gradually brightened; we could make out the shapes of the stupas and layers of mist below us among the tops of the forest. The mist peeled back and there were several volcanic peaks to be seen through the clouds. The sun peeked through the clouds on the shoulder of threatening Mount Merapi, its rays playing with the shadows below. The sky directly above us became a clear blue and we could finally appreciate the magnificent place that is Borobudur.

It is exceptional on the macro scale and the micro. You expect that it's going to be massive, but the extent and detail of the bas relief carvings going right around the temple are mind-boggling. You do experience a sense of wonderment as it unfolds in the dawn, but not religious reverence. It seems like an island of Buddhist curiosity in an ocean of Islamism. It is also a curiosity because nobody knows much about what has happened here over the last 1200 years. At some time in antiquity, Borobudur was abandoned for unknown reasons and lay half-forgotten, hidden under volcanic ash and rampant jungle for hundreds of years until it was remembered, rediscovered, excavated and cleared in the early nineteenth century.

From the bottom walking away, the temple was imposing and

Sublime Borobudur. Mount Merapi in the distance.

overwhelming. Back on the bikes, and home to Rumah Dharma for a coffee. Around 11.00 we checked out and were driven the slow 45 kilometres back to downtown Yogyakarta, to tonight's hotel: The Phoenix, another colonial masterpiece.

Yogya being the centre of Javanese culture, we had allowed ourselves two nights here. The first thing to see is the Sultan's Palace: The Kraton. The only way to get in is with a guided tour. And we were too late for today. Tomorrow then? No. Tomorrow is a public holiday. It's presidential election day. The whole country will be closed on our one day in Yogyakarta. Okay. No Kraton then. But the Water Palace, Taman Sari, was still open so we took a taxi there. What a depressing, uninteresting ruin, with a fetid pool of water and the sad remains of some odd Indo-Portuguese architecture from the eighteenth century.

We discovered a kopi luwak shop and ventured in. Kopi luwak is the famous brew in which coffee cherries are eaten by civet cats before being collected at the other end, washed (hopefully) and the beans inside extracted and roasted. I certainly had to try this! The shop lady was very welcoming and friendly. We were invited to sit at a low table in a rather spartan room and after a while an aluminium stovetop espresso pot was brought in. Well, the coffee was different: quite smooth, no acid or tannin. Very good, but hardly worth travelling around the world for. And NZ$7 for the pot, a third of a day's pay for an Ijen sulphur carrier.

Out into the street again, with becak drivers bothering us. Soon it started to spit, which was the prelude to a deluge. With only a small umbrella between us, suddenly one of those annoying becaks seemed like an excellent idea. A quick twitch of a finger and a becak spluttered across the road in a blue cloud of two-stroke smoke. We climbed onto the forward-facing seat between the wheels; behind us was the back half of a small motorbike and the smiling driver. He enshrouded us in a large clear plastic sheet to keep the rain off, and away we went. Indonesian traffic is generally a survival-of-the-toughest scenario, and our little rusting steel-tube structure was not a strong bargaining position to be in, especially on the congested multi-lane main roads that took us home, but we made it back to the Phoenix.

We had plans to meet up with Anne's cousin Dirk for the evening. Dirk lives and works in Jakarta, and today he was visiting Yogyakarta. Dirk's driver collected us and took us to his hotel, the Hyatt Regency. This has a rather more upmarket clientele than the Phoenix, so vehicle security is extraordinary: armed policemen poking their heads through windows, and the boot, and mirrors thrust under the car. Once through all that, we found Dirk and his wife, then headed out to a restaurant for a wonderful meal of fish, prawns and crabs, sitting outside under cover in the dark warmth. Afterwards, on the way home, it was interesting to see large groups of young people sitting on the footpaths, apparently hanging out in the dark. All quite placid, no music or dancing, just sitting around talking. In the absence of night clubs and bars, this seems to be what you do after dark if you're young in Yogyakarta.

Wednesday, 17 April

Yogyakarta

There was no hurry today, as we knew most of the city would be closed for the presidential election holiday. Breakfast in our Phoenix Hotel restaurant was just exceptional. The restaurant is in a beautiful indoor-outdoor courtyard setting with kitschy colonial artefacts like a grand piano and a 78 gramophone with a big horn speaker. This was topped off with an odd reproduction of the Pharaoh Tutankhamen's gold throne, slightly out of place. I've seen one before, at the Gran Lisboa Casino in Macau. Perhaps they're all over South East Asia. Here at the Phoenix, for a spot of authenticity, we had a couple of guys gently bonging away on a gamelan, lovely and very soothing if you chose to listen. There were numerous breakfast stations serving up every imaginable style of breakfast. The Yogyakarta specialties lady had an assortment of wet curries. I had her Yogya kopi which was just strong coffee boiled in a pot and served in an enamel mug. The noodle man just behind her made me an excellent mee goreng on the spot with fresh flat rice noodles, and that was me set up.

Time to hit the shuttered streets of the closed-down city. Parked outside the front door as part of the Phoenix's olde-worlde image was a beautiful cream-coloured 1950s' Mercedes-Benz sedan. I had considered buying one of these when I was 21; it was ancient even then. This one looked wonderful from a distance, but close up it was tatty and original rather than restored.

It was not for us, so we took a taxi to the Winotosastro batik factory and shop which, being a place principally to make money out of tourists, hadn't closed for the election day. It was quite fascinating, watching the ladies painting fabric with hot wax, and there was some absolutely beautiful fabric. A few young Europeans were apparently immersing themselves in batik-making courses. We decided to walk home, quite a distance in the stifling heat. After a week in Indonesia, it was exciting to be experiencing the chaotic footpaths of a big Indonesian city, even if everything was closed.

Kali Code, Yogyakarta.

Actually, after midday, some shops started to lift their shutters, and I was ready for some lunch. A hole in the wall *warung* selling soto sulung presented itself. There was only one other customer. I had a delicious bowl of soto: beef stock with brisket, for NZ$1.80, cooked on the spot in the standard Asian food cart on wheels drawn up on the footpath at the front of the shop. Anne was uncomfortable with the whole business; she ate nothing at all. After that, two or three doors further along, there was another similar place displaying the most beautiful-looking fresh, thin wheat noodles which, I convinced Anne, couldn't be all that bad. She reluctantly agreed and the man quickly knocked up a mie ayam — chicken noodle soup — for her, and she loved it.

Thus fortified, we continued our way home, getting off the main thoroughfares and walking along narrow scooters-only paths known as *gangs* where residential life was happening on the street. Away from the tourist infrastructure, people were surprisingly free with smiles and friendliness. It was all very cramped and Yogyakarta being a hilly place, there were lots of long views of how the millions live, and onion-domed mosques at every vantage point. At some point we crossed the Kali Code, the Big River In Town which looks like a sewer with thousands of houses backing onto it. We eventually made it back to the comfort of the Phoenix. The election had almost sabotaged the day for us, but we managed to snatch victory from the jaws of defeat.

We were also left with a little vignette to treasure: a young Indonesian couple enjoying the colonial columned walkways in the hotel. She started posing while he took photos. It became a chaste little dance while she chanted, 'Jokowi, Jokowi, Joko Wido-do'. He won the election, by the way. I just can't imagine anyone in New Zealand on election day doing a private little dance for their preferred prime minister.

Something I had really looked forward to in Yogyakarta was a Javanese cultural performance, the city being the centre of Javanese culture. Well, the election almost put paid to that too. But the front desk said there was one thing on tonight, the Ramayama Ballet. Brilliant. A car took us to what turned out to be the Ramayama Hotel, which had its own ballet show. Dinner was included, in the sheltered and roomy outdoor restaurant. As usual, only one other table was occupied; it was a Russian mother-and-daughter team, I think. We were entertained very quietly by a gamelan trio plus two lady singers. It was beautiful, with the tropical garden all around. A plump butterduck of a woman danced a graceful number. She had a rather masculine face and of course heaps of stage make-up, so I thought she might be a transvestite. But, no. There's no overt sex in Indonesia anyway, and certainly not same-sex sex. The food, an Indonesian buffet, was quite ordinary, but I had to try everything and ate too much.

After about one and a half hours of dinner and gamelan, we were ushered into a big concrete auditorium with a wooden roof. The place could have held several hundred people, but we were an audience of five including someone's guide. The show got under way, a gamelan of about eight players and a couple of singers. Our ballet was a complex love story involving an ugly angry stamping king, a couple of handsome knights and a few gorgeous damsels. And a chorus of monkeys including a few little kids who sometimes didn't seem to know what they should be doing. There was some very accurate archery, and a knight who had to simulate cutting off his penis. Definitely no sex in Indonesia for him. About five or ten minutes into the show, a whole busload of Chinese tourists turned up noisily and so we weren't so lonely. It was all quite entertaining for one and a half hours.

Thursday, 18 April

Yogyakarta to Jakarta

Up early, and back to Tugu Yogyakarta Stasiun to catch Train 7, the Argo Lawu to Jakarta Gambir. Seven and a half hours to cover 522 kilometres. Tugu Stasiun had plenty of people waiting, but it was all quite calm and orderly. The train journey was basically more of the same. Endless flooded paddy fields, endless houses and backyards and rubbish. Hazy skies, nothing visible in the distance. Lots of people, though. You really see people from Indonesian trains, unlike empty Australia.

The Javanese railway infrastructure gives a very good impression. There seems to be lots of new-looking track work around the stations and plenty of ongoing maintenance. The passenger trains are well patronized without being overcrowded, and the service is good. The trains are reasonably fast. They always start the day on time, though the timetable becomes a bit of a lottery later in the day.

After a brief halt at Bandung we were finally sucked into the megalopolis of Jakarta; there were several city stops before we got into Gambir in the centre, where cousin Dirk was waiting to take us to his home for the night. We piled ourselves and luggage into his van and started the crawl out of the central city. Of course, Dirk wasn't driving; there's a driver for that. Central Jakarta is massive and congested with big avenues and modern high rises, and it took us one and a half hours to get to Dirk's place. His home was a grand open-plan house with a large mezzanine with bedrooms off. It provided comfortable, spacious living in a city where space must be a precious commodity.

Soon after arriving, we sat down to a big semi-Indonesian dinner with most of the family: Dirk, his wife and two of their three daughters — good food and pleasant company. Dirk offered to be our guide for tomorrow, which promised to be fun for him also; after years of living in Jakarta it seemed he hadn't seen much of it. The evening finished with the younger daughter

Down by the station early in the morning. Tugu Yogyakarta.

breaking off the sharp end of a pencil lead inside the palm of her hand. After much digging and screaming, she and the parents went off to a clinic to deal with it while we crashed into bed upstairs in another daughter's vacated room.

Good Friday, 19 April

Jakarta

Muslim Jakarta. The perfect place to be for Easter. Tragically, not a hot cross bun in sight. We climbed into the van, with Dirk driving, for our one day of Jakarta sightseeing. Tonight we would be on a ship to Kijang on the island of Bintan.

We got ourselves to Kota, the Indonesian name for the old port of Batavia, the swampy disease-ridden capital of the Dutch East Indies from 1602 to 1942. It's where Abel Tasman sailed from in the *Heemskerck* in 1642, on his voyage of discovery to New Zealand and Tasmania. It's where Tupaia, the genius Tahitian navigator who joined James Cook on the *Endeavour*, died in 1770. It's also where William Bligh ended up in 1789 after his spectacular 6700-kilometre voyage in an open boat, following the mutiny on his ship the *Bounty*. That voyage claimed the life of one crew member, but pestilential Batavia claimed several more after they got there.

There were lots of Dutch colonial buildings all around, mostly badly dilapidated, but a few focal points were very respectably and respectfully restored. We started with a look in the Jakarta History Museum located in the 300-year-old Stadhuis. The two-storey building was fascinating but the exhibits less so. The main exhibition on show was about Australia, which wasn't really what I'd come for.

We went out and had a good look around Taman Fatahillah Square, the centre of old Batavia, and the surrounding streets. More Asian life spilling out onto the roads from falling-down Dutch-era buildings. We eventually made our way to Sunda Kelapa, the historic port with a long wharf that was the mooring place for the most extraordinary flotilla of old wooden freight ships. They were all similar, maybe 30 metres long with a very high prow and a short mast on the foredeck. The entire history of the Indonesian archipelago seemed to be expressed in those massive cracked, carvel-planked prows with layers of flaky paint splitting off them. From my perspective it was a forgotten romance of the sea, although the people on the boats didn't seem to be getting much of the romance. The odd crew member sat fishing and a few trucks unloaded bales of whatever.

We headed back towards Taman Fatahillah to escape the numbing heat. The sky went black, and next minute we were completely soaked. No raincoats of course, but at 35 degrees it didn't really matter. We poured ourselves into Café Batavia, a pleasant tropical colonial remnant on the Square. The air conditioning gradually dried our clothes and us. And the food was fabulous.

It was time to head for home; it would take a while. Sure enough, Dirk, usually having a driver at his disposal, didn't know Jakarta well enough to navigate himself. No problem; he used his phone, but it turned out he went the wrong way anyway. Then the phone battery died. That's okay, there's a cable for the phone in the glovebox. Whoops, it's the wrong cable. Suddenly we were completely lost in the city of 11 million. So we used my phone, and Apple Maps, on 3G. Very expensive, but it got us home and Jakarta is a fascinating place to get lost in. This was more exciting than any guided bus tour.

Back home, we packed and hung out with family. The daughter with the pencilled-in hand was more subdued than yesterday, and everyone was short on sleep, but no lasting damage had been done. As darkness came on, Dirk barbecued some steaks out in the courtyard next to the pool. I went out to 'help' and immediately confronted the European fear of mosquitoes in the dusk. I certainly didn't want to go the same way as all those seventeenth-century Dutchmen serving in a pestilential swamp. It seems the malaria risk is still very real. I covered myself from head to toe and then was too hot, so I went inside again.

Suddenly it was all over; it was time to go. Tonight, we would be boarding the KM *Kelud*, built in 1998 in Germany, for the 1000-kilometre, 30-hour voyage to Kijang on the island of Bintan, very close to Singapore but still in Indonesia. In planning this adventure, getting from Java to the Malay Peninsula had always been a problem. No railway comes close to covering the route. When I discovered that there was a weekly ship from Jakarta almost to Singapore, it seemed too good to be true. The ship is operated by P.T. Pelni, the huge government-owned national shipping line. The *Kelud* sails at one minute to midnight every Friday from the Jakarta port of Tanjung Priok, arriving at Kijang on Bintan at 6 am on Sunday. You can get a taxi at Kijang for the 25-kilometre trip across Bintan to the city of Tanjung Pinang, where there are fast ferries to Singapore. This just allows time, if you're lucky, to carry on to Malacca in Malaysia, all in one day. You can only book for the

Sunda Kelapa, Jakarta.

Kelud less than a month in advance. And to make it exclusive, you can't book directly and pay, from overseas. We were going to need help here.

A year previously I had got onto Dirk, asking for a local travel agent who would organise domestic travel for foreigners. He put me onto Bu Septie at VIP Travel in Jakarta who said she'd make the booking closer to the time. A week before we left New Zealand, Bu Septie made the booking and delivered the tickets to Dirk's office in Jakarta. Very convenient. And Anne WhatsApped Shandy at Bintan Taxi. Yes, he could get us from Kijang to Tanjung Pinang at seven in the morning on Easter Sunday for NZ$36. Brilliant.

'See you tomorrow,' he said.

'No, it's a month away!'

'Oh. Sorry. Of course.'

Now here we were after dinner at Dirk's place. The rain poured down and a taxi came to take us to the Pelni terminal at Tanjung Priok. Fond farewells to Dirk and all his family, and off we went into the tropical rainstorm. The port, when we got there, was huge, but we were dropped in exactly the right place. Getting through security and ticket checking was remarkably easy and next minute we were on the wharf with the *Kelud* towering over us, lit up in the dark. After a year of planning, here we were standing next to the *Kelud* with tickets in our hands.

Boarding takes you through the below-decks dormitories, large areas packed with three-level bunks, plastic-covered mattresses and no bedding supplied — just like a gigantic tramping hut. At this stage there weren't many people. A steward took us upstairs to the first-class cabins: quite roomy, with a window, two beds and ensuite. The beds were comfortable, sheets clean, but that was about it. It was all a little grimy, the toilet a bit smelly, and there were plenty of cockroaches. All pretty much what I had expected, no actual disappointment. Hey! We were on the *Kelud*!

It had been a long exhausting day and we crashed around 10.30. Loud guttural Bahasa Indonesia announcements on the in-room PA (no volume control or on–off switch) lulled us to sleep. There was no question of staying awake for the midnight departure.

Saturday, 20 April

At sea

We were woken very early by the call to prayer, loud and clear on the crackling speaker just above my head. There seemed to be just featureless sea outside our window. We were in the first-class dining room, a large clean, comfortable-enough space with plenty of morning light, by 7 am for breakfast. A trickle of locals came in with us, but the place was almost empty. Breakfast, included in the ticket price like all the meals, was self-serve mee goreng with a cold fried egg and sweet black coffee. Of the hoi polloi who spent the night in the massed dormitory, there was no sign. The day stretched out ahead of us with good books to read and an accumulated lack of sleep to compensate for. The empty sea suggested that sightseeing was certainly not going to be on the agenda. But people-watching was, as we ventured out onto the decks.

The ship was full without being overcrowded. People were gathered in small groups on benches or by the railings, smoking, looking and talking. All sorts: children, young, old, male, female. All apparently Indonesian and all at various levels of poverty. Most men were in T-shirt and jeans, a few more in traditional Muslim garb. All women wore hijabs, and while a few were in shirt and jeans they always had a hijab on top. No niqabs or burkas, though.

There was one other European couple somewhere, and we only ever saw them in the dining room; otherwise, we were the only obvious foreigners. Just about every man took at least a second look at us. A few men spoke to me in reasonable English, always inquisitive, polite and friendly. A group of men pointed us out to someone who appeared to be a clerical figure with them; he seemed somewhat dismissive or worse, but that may be my own prejudices speaking. Men didn't address Anne and no women addressed Anne or me. This was quite a contrast with the ladies' day out at the Belawan Hot Springs a week before.

The grey, calm open sea was gradually filled in with hazy smoky islands in the middle distance and extraordinary rickety fishing platforms in the middle

The *Kelud*, South China Sea.

of nowhere. It was not hot and there was quite a lot of rain, on and off. After lunch, a little sleep was in order, but it was very restless with constant public-address messages in unintelligible Bahasa Indonesia relayed to our cabin. One message which woke us up completely was particularly insistent and repetitive. We eventually got up and went out for a wander on the deck to see if something was happening. There was no one to be seen. The *Kelud* had turned into a ghost-ship. We went up another level and there was the entire population of the ship, some running around with lifejackets and hard hats. We had just missed the Compulsory Safety Drill!

As the day wore on the sea glared like stainless steel in the sun, then we were treated to a quick and spectacular sunset, thanks in part to all the smoke in the atmosphere. Dinner was an interesting low point: soup, cold fried chicken, cold fried tofu, cold mackerel heads and (cold) rice. For the next morning our scheduled arrival time at Kijang was 6 am. There were signs up outside the (closed) purser's desk saying arrival would be at 7 am. I asked two different crew members, who both confirmed that we would dock in Kijang at seven o'clock. We set an alarm for 5.30 and went to bed early.

Easter Sunday, 21 April

Bintan, Singapore and Malacca

During the night we crossed the equator, into the northern hemisphere. We woke at 5.30 am with the alarm and could immediately see land and lights moving past very close. We were in the narrow channel approaching Kijang, about to dock. It was a 6 am arrival after all! We panicked our way through the entire getting-up process minus showers and were out on the Kijang wharf soon after six o'clock, and there was Shandy's man Yogi with a big 'Gregory Hill' sign. We were out of Kijang by 6.30, heading west to Tanjung Pinang on the other side of Bintan. The road was narrow and winding, semi-rural rather than wilderness: little houses, chickens in the middle of the road, squatting Indonesians watching you go by, that sort of thing.

Soon we were in Tanjung Pinang, quite an attractive small city. We saw a lot of it because our chosen entrance to the port was blocked off by police for some reason, and we had to make a big detour through the suburbs. It was now 7.20, and there was a Majestic Fast Ferry to Singapore at 7.30! The ticket-desk people were very helpful, and the ferry even waited a minute for us. And suddenly, after eleven days in Indonesia, that was it. Goodbye Indonesia. A land of rubbish, poverty, beauty and really friendly people.

Having stumbled onto our Majestic Rocketship almost as the gangplank was lifted, a gratis two-minute noodles pottle each was thrust into our hands, and we found a couple of seats. We sailed past expensive-looking resorts, presumably for the Singaporean/Malaysian market, accelerated to a cruising speed of something like 30 knots, and the last of Indonesia fell behind us in a cloud of spray.

Time to put the clocks forward one hour to Singapore time. The trip to Tanah Merah in Singapore, next to Changi Airport, took one hour 40 minutes, and a police launch with machineguns mounted on it came out to guide us in. At Tanah Merah we had a very slow immigration queue; as it happened, this was a harbinger of things to come.

We had previously booked bus tickets out of Singapore to Malacca in Malaysia, which we didn't intend to use, just to satisfy Singapore immigration. Our plan was to take the shuttle train across the old causeway from Singapore to Johor Bahru in Malaysia, then a train from Johor Bahru to Malacca. Except the train doesn't go right to Malacca; we'd have to finish up with a 40-kilometre taxi ride, getting us in to Malacca at about 10.30 at night. Or we could cheat and just use the pre-purchased bus tickets for a four-hour trip straight to Malacca. The obvious first step was right outside the immigration hall, a bus to nearby Changi Airport where we could catch a train into the city.

From a time point of view, the convenient and comfortable bus to Malacca was starting to look better and better. So down we went to the familiar basement of the airport terminal to get an MRT train to the city centre. One change to Bugis, right where we wanted to be, one hour away. At least we caught one train today.

We surfaced at Bugis and headed towards the international bus terminal in merciless heat. It was supposedly only 32 degrees, but it felt much hotter. We sought refuge in an air-conditioned Arab restaurant for lunch. Certainly, back to real-world prices after the Indonesian experience. We made it last as long as possible then moved on to the bus terminal to wait for ages without shelter in the sun for the 3.30 bus to Malacca, a 240-kilometre journey.

Our 707 Inc bus finally turned up. Three seats across and very comfortable.

Jonker Night Market, Malacca.

But there was no time to get comfortable; very soon we were at the Singapore border control at Tuas for more queueing. Then back on the bus, on the big bridge over the Johor Strait into Malaysia where we immediately ran into the maddening brick wall of Malaysian immigration. What a debacle! We had to stand in the queue for an hour, inching forward. The locals, Singaporean and Malaysian, seemed inured to it, but you had to keep your eyes out for opportunistic Chinese tourists trying to push their way through. Nobody was enjoying themselves, least of all the bored, tired immigration clerks. Once we got to the head of the queue, it was all straightforward and we were back to the bus to wait for those who were even slower than us.

When we got going again, it was on a modern fast motorway towards

Malacca. This was a completely different place than Indonesia: no rubbish and no obvious poverty. At least around here, there was no rice growing, just an endless monoculture of palm oil trees. We finally made it to Malacca in the dark and took a taxi to the magnificent old Majestic Malacca Hotel, checking in around eight o'clock with the full five-star treatment of tea, juice and lovely little sweetmeats. At the end of a long day spent travelling through three countries, it was most welcome.

The best news was that the famous Jonker Night Market was on. So we charged up to our big airy room, scoffed the little sweet things, then headed out on foot to Jonker Street, a 20-minute walk away. What a riot of light, colour, noise, humanity and food. In the middle of the street was a big ornate stage looking like a particularly gaudy Chinese temple. A big crowd, mostly elderly, sat on standard Asian plastic chairs enjoying a solo performance by a septuagenarian male rock-and-roller with recorded accompaniment. Another very old man was on the stage shuffling and posing to the music, mostly with his back to the audience. It was all very weird, but the audience seemed to appreciate it all.

There were endless socks, underpants, teddy bears and electronics for sale in stalls, but I was only interested in the food. Food buying had to be severely rationed and disciplined, because there was so much variety but only one stomach, with a lesser interest from Anne. The deep-fried baby crabs were a crunchy must-have. And one buttery Hokkaido scallop with some greens and noodles, all in the shell, turned out to be a small meal on its own. There was only room for one Rockefeller-style oyster after that, and a Tiger beer.

We walked home the long way along the meandering Melaka River, a stinky grungy magical place. Coloured lighting on the embankments and little bridges brightened the place as we strolled past small family houses, bars and godowns. This was pretty much the Malaysia I was looking for. We finally made it back to the Majestic Malacca about 11 pm and were asleep within seconds.

Monday, 22 April

Malacca

Out into the blasting heat and noise, ready to discover Malacca by daylight. What a melting pot! Chinese immigrants started arriving 600 years ago; they intermarried with the indigenous Malays and stayed. Then the European colonizers took over, 500 years ago: first the Portuguese, then the Dutch and finally the British, who left in 1957. New Zealand fought a war here, alongside the British, in the 1950s. The Malayan Emergency was a successful attempt to stop the communist spread. And everybody has left their mark on the place, even the ethnic Malays who were here before the Chinese, and who, I believe, now number less than half the population.

Malacca is no longer the trading hub it once was; with its UNESCO status and busloads of Singaporean tourists, the centre of town is more like a functioning museum. We set out on foot into the centre along the embankment of the Melaka River. There were huge monitor lizards lazing in the mangroves by the river. The first one gave me a real shock as I had never heard of them, and I thought at first that it was a crocodile. But they were everywhere and soon lost their novelty status. We left the river behind and headed down streets full of warehouses and shophouses in various stages of dilapidation but most of them in use for something. And all with a probably unintentional but remarkable aesthetic quality to them. I was liking this despite the miserable heat.

We were heading to the reputed Nancy's Kitchen for lunch. We'd done without breakfast after the excesses of the night before. Besides, it's impossible to eat much in this heat. Nancy's Kitchen is a small shrine to Nonya cooking, the unique Chinese-Malay cuisine of the Straits Chinese. Right here on the Straits. You can't get much more authentic than that. Just like risotto alla Milanese in Milan or Wiener Schnitzel in Vienna. We sat in the comfortable front room of a colonial shophouse, and it looked like the other tables were all filled with local family groups. No white faces anywhere. I think the word

Malacca.

'Nonya' refers to Mama and there is a certain sense of mother-dominated family comfort in the ambience of this restaurant, and the food. Of course, we ate very well.

Replete, back out into the blinding heat, we wandered along the shophouse fronts towards the river mouth, stopping off for quenching cooling pineapple and watermelon on the way. The river mouth is on the actual Straits of Malacca, rather spoiled by a motorway overbridge. Here was the Malacca Maritime Museum, featuring a towering replica of the *Flor de la Mar*, a Portuguese carrack which sank in the Straits in 1511 with a large cargo of looted Malaccan treasure on board. I don't think the replica was particularly authentic, but it was worth a good look, and the rest of the museum was a pleasant respite from the heat outside.

We moved on to shady Bukit St Paul (St Paul's Hill) to catch up on more early European Straits history. Here the Portuguese built a large fortress in

1511, the same year the *Flor de la Mar* sank. The British demolished it 200 years ago and all that's left is one massive squat, ugly gatehouse. Walking past this we were into a quite pleasant park area, and someone had seen fit to preserve and display a rather unattractive Scottish Aviation Twin Pioneer transport aircraft which the Malaysian Air Force used in the 1960s, along with a similarly aged Denis fire engine parked next to it. It was identical to the ones seen in my childhood, in another corner of the British Empire. On then, up the steep steps of Bukit St Paul, past the buskers and trinket sellers to St Paul's Church on the top. Actually I couldn't save myself from buying a Malacca streetscape watercolour from a friendly artist under a tree. He spoke fluent English and was interesting to talk to. I carefully protected the unmounted painting in my soft travel pack for the rest of the journey around the world, and now it's hanging, framed, in our living room.

And so to the extraordinary church of St Paul. Built by the Portuguese in 1521, it's 120 years older than the more famous Portuguese St Paul's façade in Macau. There is no roof, but the structure is quite intact. It's fascinating to walk around, with sixteenth- and seventeenth-century Portuguese gravestones later mounted on the walls, and magnificent views through the windows and doorways down to the town and the Straits. When the Dutch turned up in 1641, they reconsecrated the old Catholic church as Dutch Reformed, and in 1824 the British, with their Henry VIII-inspired brand of pragmatic Protestantism, devoutly repurposed St Paul's as a gunpowder magazine. Fire and brimstone indeed. And they built an uncomplimentary lighthouse right next to it in the nineteenth century.

From Bukit St Paul we headed down the other side to the Red Square, the colonial town square with the Stadhuis, older than the Batavia one, and Christ Church, everything painted in an ochre red. This really was Tourist Central, with the most extraordinary garishly decorated trishaws complete with their own blaring pop music. It seems a ride in one of these things is an absolute must do, but we demurred. We found a quiet restaurant for dinner, obviously aimed at the Western tourist market with 1960s–70s British record covers on the wall: a bit Bohemian. I thought I should eat a laksa — spicy noodle soup — while in Malacca and it was really good.

That was us pretty much done for the day, so we wandered back to the Majestic Malacca and our room full of dangling dry laundry. I'd had a fabulous day in Malacca; I'd always wanted to see this storied old crossroads of trade and migration.

Tuesday, 23 April

Malacca to Kuala Lumpur

Out of bed at 5.45, as we'd ordered a taxi for the 40-kilometre trip to Tampin/ Pulau Sebang, to catch our first Malaysian train, to Kuala Lumpur. Tampin/ Pulau Sebang Station was very new and quite large. The driver caringly dropped us at the totally deserted wrong side of the station. This side of the place was completely closed first thing in the morning, the platforms and tracks secured behind wire fencing. The only way across would be through the locked terminal building. Through the wire we could see people starting to arrive at the main entrance on the other side. Eventually, the doors on our side were unlocked remotely and we were up in the lift, over the rails on the covered footbridge and down to the main station building.

We still had plenty of time, and there was a promising-looking kopitiam café opposite the station entrance. Anne wasn't keen but I was, and we had to eat something for breakfast. The Indian man at the counter made us a beautiful Roti Chennai breakfast for 37 NZ cents. Train ETS 9420 turned up with its purposeful bullet-train snout, on time at 8.10, and we set off on our 120-kilometre journey to Kuala Lumpur, just under two hours. It was a very comfortable train, getting easily up to 130 kph between stops, very smooth on only a 1-metre gauge. The fellow passengers in our carriage seemed to be middle-aged middle-class Malaysian men going on, or returning from, a golfing holiday.

After endless palm oil plantations, we eventually descended underground and pulled into Kuala Lumpur Sentral downstairs in the dark with no lights or signs. I didn't realize we'd arrived until everybody got off with their golf buggies. Finding a taxi was easy enough and soon we were at the Traders Hotel in the city centre. This was a fairly luxurious modern city hotel with none of the period charm we were becoming familiar with, but with a drop-dead view from our room of the unbelievable Petronas Towers. These two 88-storey skyscrapers were the tallest buildings in the world (451 metres)

when they were opened in 1999, though they've lost that badge now.

We ventured out through the lovely City Central Park and through the arcades at the base of the towers, and soon found ourselves at the entrance to the home auditorium of the Malaysian Philharmonic Orchestra; it was fascinating to look at their posters and see what was coming up in classical music in Kuala Lumpur. The musical world being what it is, Anne and I had several colleagues who had worked in that orchestra.

Jalan Alor, Kuala Lumpur.

I was trying to follow the advice of the Vietnamese-Australian food writer Luke Nguyen and find some restaurants that particularly appealed to me. We were looking for something called the Imbi Food Market on Jalan Imbi. There's supposedly a place at the market that does excellent kaya toast, a Malacca Straits breakfast of buttery white toast with sweet coconut jam and a soft-boiled egg, and milky tea. Well, after a bit of walking around in circles, several different MRT trains and a monorail, we eventually got to Jalan Imbi. We walked the whole length of Jalan Imbi in oven-like heat and I swear there was no food market there. Locals I asked didn't know what I was talking about. So no kaya toast then.

We were exhausted from the excruciating heat and had to sit down somewhere at the side of the road. A lady cooked us a delicious tom yum soup. Thai, not Malay, but pretty good anyway. Revitalized, we had the energy to walk home to the Traders. We discovered the extraordinary system of elevated walkways that Kuala Lumpur has. With them, there's no hassle of street crossings and you can cover long distances quite quickly.

Back at the Traders we met up with our friends Agnes and Ali, a German-Indonesian couple who, by the greatest of coincidences, just happened to be visiting Kuala Lumpur on the same day that we were. We had all known about this maybe a month in advance, but it was still hard to believe that we had zeroed in on Kuala Lumpur on the same day, with no need to adjust schedules.

After joyful greetings we set off on foot for Jalan Alor, a celebrated food street. After going around in more circles and walking too far we eventually found it, a pedestrianized street packed with tables served by restaurants and hawker stalls. Unlike the Malacca Jonker Street Market which was a 'walk along stuffing your face as you go' sort of place, here we could sit down, order, talk, drink and eat food off plates. We walked the length of the street first, to see which would be the best choice. Most places seemed to be basically Chinese with a bit of Malay influence, no Indian and certainly no Nonya. However, we discovered a Chinese eatery with excellent Malay-style black-pepper crab and there was no need to look further. What a fabulous evening: shirt-sleeve warmth, superb fresh food, Tiger beer, a unique catch-up with friends, and the sense of being in the middle of an extraordinary adventure.

When I'd wiped the last of the crab juice from my face, we staggered off in search of a rooftop bar that Ali knew about. This involved a lot of walking through a darkened, rather dilapidated medium-rise residential area which I found fascinating: so this is how people live in Kuala Lumpur. We finally found the right place and ended up outside on the roof, 33 storeys high. The view was stupendous. We stayed for ages, but eventually this little episode was over, and we came back down to earth and said goodbye. Agnes and Ali would be spending a few days in Kuala Lumpur, but tomorrow Anne and I would be on the train to Penang.

Wednesday, 24 April

Kuala Lumpur to George Town

The day began with a return to the included-in-the-tariff hotel breakfast and this one was seriously serious. What was on offer was extensive for every cuisine. I was determined to get my kaya toast. It wasn't on display anywhere, but a man in the Malaysian department was very happy, and somewhat surprised, to run one up for me. I think Malaysians would not bother to eat something so common in such a posh place where you can have anything you want. It was delicious and worth the fuss, but unfortunately it filled me up so much there was no room for all the other wonderful stuff.

Meanwhile, we had a train to catch, to Butterworth in Penang where the Andaman Sea funnels down into the Straits of Malacca. Butterworth was 350 kilometres away, just over four hours on the train.

At the Traders concierge desk, we asked for a taxi to Kuala Lumpur Sentral Station. The concierge said there was no need for us to take the airport train, as a taxi straight to the airport would be quicker. He couldn't understand that a couple of white tourists would want to leave Kuala Lumpur by train, and that the airport was not in our itinerary for the day. Finally, we got into a taxi and the driver said the station was not a good idea, straight to the airport would be better . . . There were signs in the taxi advising us that throwing rubbish out the window was a very bad idea and kissing was even worse. Early-morning garbage disposal and red-hot passion were put on hold then. Somehow, we ended up at Kuala Lumpur Sentral Station. The station is huge, and of course the driver dropped us at the special entrance for the airport MRT train, miles from where we wanted to be. So we walked until we found the long-distance terminal, a massive modern place with crowds of busy people coming and going. How come nobody on the outside knew it was here? But we found our train, ETS 9102, similar to yesterday's, and left Kuala Lumpur on time at 11.30 am.

Kuala Lumpur was a fascinating place. We didn't even scratch the surface and it's a pity we didn't spend more time there; somehow it didn't raise the

Penang Strait.
Approaching George Town.

'must-see' flag when I was planning. When you're travelling around the world over land there just isn't time to stop and see everything. Being newly retired, I had plenty of time, but Anne had to take leave without pay from her job, so it was quite expensive for her.

Now here we were leaving Kuala Lumpur; within a few minutes we were travelling through the old 1910 Kuala Lumpur Station, no longer used as a long-distance terminal. It's an extraordinarily grandiose imperial place, a cross between a Moorish palace and a wedding cake. Oddly, a train is not the best vantage point for appreciating a station and I'm sorry I missed most of it. We got out of town and up to our smooth cruising speed of 130 kph. I don't know how they do it on a 1-metre gauge; the prehistoric New Zealand trains struggle to manage 80 kph on a 1.067-metre gauge. Most of the journey was uneventful and uninteresting. No rice paddies, just the usual palm oil trees. There was some hazy coastal scenery near Butterworth. The train was full, and you have to pre-book; you can't just turn up on the day expecting to travel. We arrived in Butterworth one minute late.

For us and most others, Butterworth was just a transit point. We were here for the 20-minute ferry trip across to Penang Island, to George Town, the capital of Penang state. The ferry terminal was a simple walk from Butterworth station: crowds of people, including a few white faces, queueing up to buy NZ40-cent tickets for the ferry which leaves every 20 minutes. The ferries are of gorgeous old two-level open design, drive-on and -off for cars and with room for pedestrians to sit and stand among the cars and around the edges. They are at least as special as the Hong Kong Star Ferry, and George Town looms up out of the tropical sea mist just like a mini Hong Kong. On the other side we grabbed a taxi with an extraordinary tiny wizened old woman driver who just would not stop talking. She took us to the Eastern and Oriental Hotel on the waterfront and by the time we got there I was an exhausted wreck from trying to keep up with the assault of her rasping monologue. Don't get the wrong impression, she was lovely, but like a bored cockatoo in a cage.

Arrival at the Eastern and Oriental tipped us into a world of classical opulence. Our room was big with a fabulous sea view north-west along the coast. As usual, we were out walking immediately, looking for dinner ideas. We ended up at a little place on a street corner, eating kung pao octopus (yum), char kway cheow and lamb satays. And a Tiger beer of course. Then home to bed.

Thursday, 25 April

George Town

As the day hadn't warmed up yet, we breakfasted outside in the Eastern and Oriental's shady garden next to the sea. It was astoundingly beautiful and peaceful, with the sun struggling through the clearing mist. Nobody else ventured outside; we had the whole place to ourselves. Mad dogs and Englishmen — and Kiwis.

After this we hit the streets and stopped off at the locally famous Blue Mansion, more correctly known as the Cheong Fatt Tze Mansion, after the self-made man who had it built 140 years ago. It's a very rare surviving example of how the rich Straits Chinese merchants lived in a previous era. More than just a huge house, it is extraordinarily decorated with ceramics and painted panels and full of old Asian bric-a-brac. It seems the then newly restored mansion was used as a location in the 1992 movie *Indochine*, and lots of props were brought from Hanoi to make it look authentically Vietnamese. So we saw a collection of old Chinese string instruments, erhu and huqin, which presumably have Vietnamese names I'm not familiar with, and fabulous old Hanoi rickshaws, man-powered, one passenger, big wheels. You don't see anything like that any more.

We were ready for some lunch and today's project was to eat Assam laksa in Penang. It's very different from the coconut-flavoured laksa you find further south. We didn't see anyone offering it, so we sat down at a corner footpath place for a snack of something else. The woman in charge sat down and spoke with us. She was obviously chuffed with our enthusiasm for her food, and we told her about the Malaysian restaurant scene in Wellington. She said that for Assam laksa we must go to an area called Air Itam, 8 kilometres away, and told us what bus to catch, how much to pay, when to get off. So we slurped down the last of her noodles and found ourselves on a bus through the suburbs of George Town.

Air Itam is right on the outskirts, at the foot of the hills and full of Buddhist temples and pagodas. Sure enough, we found a rough shady garage-like space

Air Itam, George Town. Kek Lok Si Temple in the background.

with no customers where the man cooked me a fabulous Assam laksa, an explosion of minty tamarindy flavours, while the restaurant cat entertained us. We were at the foot of the Kek Lok Si Temple complex which stretched up the hill above us, so we ventured in. Past a big pond full of socializing turtles which were hard to get away from, the path took us up and up through ornate Buddhist pagodas and lots of tourist shops. This really was commercialized Buddhism on a massive scale. Halfway up we abandoned climbing and took a steep cable car instead. There were beautiful formal Chinese gardens with the forested hills for a backdrop, long views down to the distant city and the sea and ranks of kitschy gold-clad pink Buddhas decorated with slightly confronting golden swastikas, which apparently symbolise the auspicious footprints of the Buddha. At the top is a 35-metre-high bronze statue of Kuan

Yin, the Chinese Buddhist goddess of mercy. Back down to the bottom, and an interminable wait in the heat for a bus back to town.

By now we had run out of Malaysian ringgits and we needed an ATM so we could continue to eat. After some difficulty we found one, but it wouldn't work for us. We walked around and tried a few more with no luck. They all demanded a registration number which we didn't have. It seems that all the Malaysian-owned banks would only accept Malaysian cards. I've never come across anything like that anywhere else in the world. My ASB debit card has always worked everywhere from Invercargill to Iceland. But the Malaysians do things differently and so we decided to try a foreign-owned bank. After miles of hot thirsty walking, we stumbled upon a very upmarket and massive HSBC Bank. Of course, I pressed the wrong button on the ATM, and it swallowed my card and would only ask me questions in Arabic. So I froze, not wanting to make things worse, and eventually it ran out of patience with the stupid infidel and desultorily spat my card out again. I tried once more, carefully selected the English option, and finally I had a fist full of ringgits to fold into my wallet.

Nearby was the old British defensive structure of Fort Cornwallis, built at the beginning of the nineteenth century, situated next to the sea, defined by massive stone walls and an empty moat. The fort never fired a shot in anger, though interestingly enough the German U-boat that reconnoitred Gisborne harbour in New Zealand in 1945 was stationed for a while at the Japanese naval base right next to the walls of Fort Cornwallis.

One feature of the fort is a huge seventeenth-century cannon made in the Netherlands. The Dutch gave it to the Malays as payment for trading concessions, then the British took it by force, and now it rather decoratively stands sentry over the Penang Strait at Fort Cornwallis. It has some beautiful details on the bronze-cast barrel, including a perfect rendition of the distinctive Dutch East India Company logo, which has always fascinated me. It's a stylisation of the letters VOC, for Vereenigde Oostindische Compagnie. Designed in the seventeenth century, is this the world's first corporate logo?

Enough of military and corporate history, it was time to eat again. Apart from breakfast at Tampin/Pulau Sebang, we hadn't eaten any Straits Indian food on the Straits, so we opted for dinner on the footpath at an Indian hole in the wall. Beautiful tandoori ayam, that's chicken, for Anne, and murtabak, a pancake stuffed with meat and smothered in curry, for me. Ridiculously cheap. On the way home we found a bar and stopped off for a beer. Ridiculously expensive. In our now-usual evening state of exhaustion, we staggered home for our last night of opulent Eastern and Oriental Luxury.

Friday, 26 April

George Town to Hat Yai and beyond

Further into the unknown: a big day with a ferry, three trains and the border crossing into Thailand. The Thai border, in particular, was flagged with foreboding on the internet. We checked out and left earlier than we needed to, as I was nervous and wanted to get going. The ferry back to Butterworth was free, included in the 40-cent outward fare.

We took a KOM commuter train from Butterworth to the border at Padang Besar, 166 kilometres north, a two-hour moderately scenic trip in a spotlessly clean, reasonably comfortable commuter train. The steel seat-backs made it look from behind like a toilet carriage with 40 urinals in it. The train was quite full, with a very Muslim clientele. At one stage there was a rather military-looking police convoy travelling along the highway beside us. It really felt like we were getting further away from comfort as we approached the north end of Malaysia and the Thai border.

Padang Besar seems to exist only because of the border and the necessary railway infrastructure, and of course it's very hot. Probably every day of the year. We got off the train and wandered around the empty Malaysian and Thai immigration halls, which fortunately at this border were in the same building. Eventually, a cleaning lady told me where we should be, and a Malaysian official emerged from a stuffy, smelly, smoky office and said that the train would come in two hours and immigration would open then. We were waiting for a Thai local train for the 43-kilometre trip to Hat Yai. At Hat Yai we would wait a couple of hours then take train SP 38, an overnight sleeper service all the way to Bangkok.

After a while it became apparent that we could now go to the Thai area of the station to buy tickets for the train to Hat Yai. So we left the comfort of air conditioning and queued outside in the sun for quite a while; the ticket office only had room for about the front five people in the queue. Out the other side, and everyone congregated in the Thai immigration area to get out of the sun. Sometime later we were all told to clear the immigration areas and wait

MYANMAR
LAOS
THAILAND
Bangkok
CAMBODIA
Aranyaprathet
Poipet
Siem Reap
Tonle Sap
Santuk
Phnom Penh
Phuket
Hat Yai
Padang Besar
George Town
MALAYSIA
Rail
Sea
Road

outside. There must have been a couple of hundred of us or more.

The train from Hat Yai, our train, wheezed and clanked in to the other side of the platform, decked out in its distinctive Thai Railways purple, white and yellow. The people on that train had full call on the Thai and Malaysian immigration facilities until they'd all been cleared. They came smiling out of Malaysian immigration in depressingly slow dribs and drabs, while we desiccated in the sun.

An hour or so later we were finally allowed to have a go ourselves. Our crowd turned into a queue for Malaysian emigration; it took a while but went smoothly enough. Out a door and in another, to Thai immigration. Miraculously we were directed to a queue for foreigners, which consisted of just Anne and me. That was very quick and we hopped over to the other side of the platform, still in Malaysia, and onto our Thai train. This was a proper old-style Asian train with minimal comfort, open windows and broken seats. It was ancient. There was still waiting around to do, as people were only trickling out of Thai immigration. Thank goodness for the foreigners' queue!

We clattered out of Padang Besar and Malaysia at five o'clock, an hour and a quarter late, after six days in Malaysia, our first late departure in Asia. Into Thailand and the clocks went back one hour to Thai time. The journey was only three quarters of an hour anyway, so we arrived in Hat Yai before we left Padang Besar. And it was a really enjoyable train trip. With the hot wind blowing through the windows and through my hair, I felt I was in the landscape, with all its noises and smells too, and it was slow enough to count each blade of grass. We had definitely left rich Malaysia and were back in poor Asia. Rice paddies again, cows, banana palms, not a palm oil tree to be seen. After a 45-minute riot of colour, noise, wind and life we passed through a graveyard of derelict and rotting diesel locomotives and pulled into Hat Yai. The station consisted of a long main platform with several other platforms, all accessed straight over the rails.

Our first job was to find the Parcels Office, to collect our pre-booked tickets for the train to Bangkok. Not so easy, for while there were signs around the place, they were all in Thai script. We found one useful sign in English: 'Information'. The information girl spoke good English and was pretty smart; she led us on a long walk to the end of the platform to the dingy Parcels Office. Our girl explained to the woman there that we were here to pick up tickets. We had booked through a Vietnamese company called Baolau, which seems to be the only way for a foreigner to book this train and pay with their foreign credit card. The Parcels Office lady was obviously familiar with the

Clattering into Hat Yai.

situation. She reached over and pulled off the shelf a large stack of big, stiff brown cardboard envelopes. She went through them one by one, but couldn't find ours. Time was starting to tighten up, the train would leave in half an hour, and our tickets were apparently not here. The information girl was starting to tell me I should contact the Baolau booking agency.

I was imagining finding Baolau's phone number — did I even have it? — in an email and trying to phone them in Vietnam from Thailand. The heat was becoming oppressive. I asked if the Parcels Office lady would look through the pile again. Please. She was happy to oblige. She quickly shuffled through them. I watched this time. All the names were upside down from her point of view. I saw a flicker of 'Gregory Hill' and yelled out 'There it is!' I dived in and rescued our tickets. It appeared she couldn't read Latin script and didn't know it was upside down.

Now was not a time for recriminations, just relief. The Thai signs outside looked like beautifully jumbled nonsense to me, so I sympathized with her inability with English names. Goal achieved, tickets to Bangkok in my hand, we said sincere thank-yous to the helpful information girl, and trundled back to the big airy waiting room where we found out that the train would be half an hour or so late, and we discovered we were starving. We had had no food since breakfast at the Eastern and Oriental in a previous era. So I popped out to one of the many stalls on the platform and got us a couple of pieces of unadorned crispy fried chicken and a large bottle of water. That did the trick.

After a while the newly appointed arrival time for our train approached and we made our way over the rails to the correct platform. Sure enough, a big, noisy red-orange-and-yellow locomotive growled in, pulling a uniformly motley collection of dirty, greasy, rusty carriages. The so-called platform was at rail level so the whole thing towered darkly above us. Our sleeper tickets were first class and there was only one first-class carriage, at the front directly behind the guard's-van-type-thing which looked like the interior of an unkempt suburban garage. We dragged our packs up from ground level into our carriage and entered a dilapidated, dirty, narrow passageway. We found our cabin number and entered our cell for the night: very small, quite unpleasant and unclean — and we would have to spend sixteen hours in here! I had been hoping for a bit of luxury at Thai prices. We left around 6.30 pm, half an hour late, still on the narrow 1-metre gauge, with 936 kilometres to go.

We felt rather trapped in our cell, and hungry, so a trip to the dining car was a good idea. We set out through darkened carriages full of people hanging out in the aisles with all the bed/seats obscured by long, dangling blue plastic curtains. People were eating, getting changed, carrying bowls of food; it was like a chaotic Asian street scene played out in the narrow confines of a train carriage. It was an absolute obstacle course to negotiate, but after about three carriages of this we made it to the relative spaciousness of the pale green enamel-and-filth dining car with a serving and kitchen area at one end, and grubby tables next to open windows with the dusk blasting in. There were opaquely laminated sticky menus on the tables, fortunately in English. We ended up with a couple of plastic trays of palatable Thai curry.

With Thai pop on the speakers and a few security men eating, with static and garbled voices on their walkie-talkies, and the last of the daylight flitting by, the place certainly had a definitive ambience. From time to time the most appalling sewage smells wafted in through the windows. It made eating quite difficult. We didn't linger and wended our way back through the Thai village on rails to our little cell. Someone had been in, put the beds down and made them up! The sheets were clean and the place looked quite habitable after what we had seen elsewhere on the train. It was early to bed then as we trundled through the dark, up the long southern Thai peninsula, with the train's horn sounding in the near distance about four or five times every minute.

Saturday, 27 April

Bangkok

The night was pretty rocky and noisy. Approaching Bangkok was an endless crawl through a jungle of motorway flyover supports. We pulled into the huge arched interior of Hua Lamphong Station at about a quarter to eleven in the morning, half an hour late, and there on the platform was a sight from home: the fabled Eastern and Oriental Express had arrived from Singapore. This train used to belong to New Zealand Railways in the 1970s, when it was the Silver Star luxury sleeper service running between Auckland and Wellington. It was now about 45 years old and looking quite tatty, at least on the outside, no longer the gleaming silver modern marvel that I last travelled on in perhaps 1977. I do remember that it was very comfortable, and now, as the Eastern and Oriental Express, it has a fabulous reputation for luxury service. I was very happy to see it this morning.

Now we needed to find our hotel. We did have the intention of taking a very slow train to the Cambodian border tomorrow at about 6 am, so I had booked us a cheap convenient hotel close to the station. Out through the ornately colonnaded and castellated front entrance of Hua Lamphong Station into the extreme Bangkok heat. This was quite serious. While Singapore is usually about 32 degrees, Bangkok is usually 36. We immediately saw our Prime Hotel Central Station: a tower about 100 metres away, across a profusion of roundabouts and traffic. Thai drivers are quite courteous to pedestrians, but only if you show real determination. They give you about half a metre or one second. Once we got there, our hotel room was surprisingly good: spacious, clean and with a brilliant view down to the station and beyond to the more distant high-rise part of the city.

This was not what we were here for though, as there was food to be explored! I wanted to follow up on some recommendations from my son Francis, who had been here before, and the previously mentioned Luke Nguyen. So into a tuk-tuk and off to Thipsamai, the famous pad Thai place in Maha Chai Road. Well, they were closed; they'd be opening at 6 pm. We were getting quite hungry; we'd had no breakfast yet. Drinking copious amounts of water, we walked back through Chinatown: a chaotic jumble of shark's fin joints, bird's nests, gold merchants and filth. Eventually, starving, we found a Chinese footpath place for crispy roast pork on rice, made Thai with a fish-lime-chilli sauce. Carrying on, we ended up at the food stalls near the Golden Buddha, and a place doing excellent oyster omelettes.

Now with full stomachs, we thought sitting down for a cruise on the Chao Phraya, the big river that flows through Bangkok, would be a good thing. A tuk-tuk driver delivered us to his mates and we paid 1000 baht, that's NZ$47, for a one-hour cruise. It was actually terrific. The boat was long and narrow like a Venetian gondola, but rather overpowered with a heavy car engine in the rear that really lifted the bow at a fast cruise. There might have been room for about ten people, but we had it to ourselves. No commentary; the man just drove the boat and left us to our own conclusions.

We headed out into the main stream of the wide Chao Phraya, and the water was quite turbulent from the conflicting wakes of a lot of vessels going too fast. Our boat certainly didn't feel particularly stable and we had to slow down a few times. Off the main highway then, and into the side channels, and this is where it got really interesting, being close up to all the houses — all sorts, grand and poor, upright and collapsing into the water. This was all low-rise stuff on piles in the river, left to rot into the water. The canals are the thoroughfares in this part of town, sort of a Venice of the East but without the sense of history. Eventually, after briefly entering the domestic lives of the canal-dwellers we headed out into the river again, for a precarious transit to the other side; it was maybe a couple of hundred unstable metres, or more. This got us back to Chinatown, and a little piece of fried fish on the footpath.

We walked home, which took a while. Exhausted as usual, we had a nap, then it was time to find some dinner. On a dark street corner, we found Nirvana. Crowded tables, with people contentedly slurping hotpot and barbecued meat, and an empty table for us. A bit of pointing and sign language had us sorted, and we were brought a domed steamboat-type-thing rather like a giant lemon squeezer, powered by a gas bottle, where you put thin slices of meat and fat directly onto the central raised dome, and vegetables into the water in the surrounding moat. No stock, just water; the fat and meat juices trickle down into the water and the staff are on hand with a beaten-up teapot of water for if you start to run dry. Smashingly good, with hot sour Thai dipping sauce and lots of Beer Chang. We felt victorious at our discovery, and staggered home to our roomy unmoving bed at the Prime Hotel Central Station.

Bangkok.

Sunday, 28 April

Bangkok to Siem Reap

This morning we had a bus to catch — to Cambodia. We had originally planned to take a very-early-morning train terminating at Aranyaprathet near the border, then find further transport in Cambodia. However, this option was flagged as a potential trap for the unwary, so I had also booked an international bus direct from Bangkok to Siem Reap, just in case. Now our courage deserted us; we abandoned the 6 am train and opted for the safer and later bus. Our Virak Buntham Company bus would leave from an obscure spot, marked on a map, on Tsao Fa Road. The location was a mystery to me, so we allowed plenty of time and I trusted that it would make sense to a taxi driver.

Leaving the Prime Hotel Central Station, we found a driver who assured us he knew where the Cambodia bus terminal was. I really wanted to believe in him. Our man insisted on a flat fee of 200 baht, no meter, which I thought could be a good deal in the event of him getting lost. Sure enough! I followed our progress on my off-line maps app. It couldn't navigate but showed where you were and where you wanted to get to.

We got off to a promising start, but after a while we were definitely heading in the wrong direction. In the middle of a packed eight-lane highway I told the driver he was going the wrong way, and he wasn't pleased. Major remonstrations in both directions. He was very grumpy and got me to navigate with my app. I eventually got us on the right track, but it was a long way to go now. He was sure I didn't know what I was doing and there would be no bus terminal at the end of it.

Just short of Tsao Fa Road, the app showed a huge complex spaghetti junction between us and it. I knew I couldn't negotiate this with the app, so I got the driver to stop right in the middle of the first roundabout and we got out and walked. We had to be extremely careful here, as the traffic was very heavy and going in every direction on this complex system of roundabouts, on-ramps and off-ramps. So we took our lives in our hands, dragging our packs behind us, and sprinted across several multi-lanes of automotive spaghetti. We seemed to be very close to the beginning of Tsao Fa Road.

At one stage we had to divert a little, upstream so to speak, to find a safe and predictable place to cross. Just like walking upstream to find a safe river-crossing when you're tramping. We got to a quieter spot and I looked up and there was a bus parked at the side of the road. I thought that's an odd coincidence, it's got Siem Reap written on it. Aren't we supposed to be going there today? Then I saw the company name, Virak Buntham. We were home and hosed! We had somehow stumbled upon our bus. It was sort of in the right place, although the location didn't look anything like the helpful photo on our print-at-home ticket. Stress levels had been running quite high and the feeling now was of absolute relief. We were going to leave Thailand today after all.

We seemed to be the first passengers as we climbed into the empty Cambodian bus, decked out with patterned curtains and opulently padded purple armchairs, three across. Along with the previous Malaysian bus, this was a comfort revelation after a lifetime of orchestra bus tours in cramped four-across seating. The bus gradually filled up; goodness knows how people found it. We set off ten minutes late and crawled through the Bangkok clutter . . . to Hua Lamphong Station. Just across the road from our Prime Hotel Central Station, for a scheduled pick-up stop. I couldn't believe it. When I booked this bus online, there were options for different pick-ups, but they didn't make sense to me so I thought the starting terminus would be the best bet and we could just catch a taxi there. We weren't committed to taking this bus anyway.

After leaving Hua Lamphong it took forever to get out of Bangkok, even on good flowing motorways. Once out of town the countryside was dull and flat, with the standard South East Asian haze, but the roads were still good and the Hyundai bus, which seemed very powerful, really went for it. Which was just as well because this was a 410-kilometre trip with a big border crossing in it.

We passed through Aranyaprathet, where the Thai train terminates. The bus stopped in the middle of a circus of casinos and hotels for Thai emigration. The foreigners' emigration hall was upstairs and straightforward. From there we were shepherded together and headed off in single file across The Friendship Bridge, built by Thailand with British money, over a rubbish tip and a fetid stream which is the border, and into Cambodian immigration at Poipet. There was a big queue here, but it moved quickly. The unfriendly

Siem Reap.

immigration officer apparently didn't do the right thing with my passport; after I walked through there was a commotion and I was called back. They weren't going to let Anne into the country without fixing my passport! We got all of that sorted, and so 'Welcome to Cambodia'. Now it seemed a real relief to just climb back onto the bus rather than deal with the supposed thieves and rip-off merchants in miserable heat and try to negotiate onward transport from here.

Train travel through Cambodia was not an option for us. A new railway line from Poipet to Phnom Penh had recently been completed, but there was not, as yet, a reliable or regular train service, and we needed to keep moving.

Away from Poipet, Cambodia was immediately very different than Thailand: rough roads, boring flat landscape, heaps of earthworks, unutterable poverty and rubbish, and vast areas of what looked like dried-out rice paddy. Was this the dry season? The place was desolate with sad grubby little villages and skinny cows trying to find grass between the dirt. After a few uneventful hours of this we got to Siem Reap about 5.30, past an extraordinary line-up of kitschy hotels on the main road out to Angkor Wat. Somebody's making plenty of money here, just not most of the locals.

We were here sixteen years before; most of these places didn't exist then and I barely recognized the road. We went to the stunning Angkor Wat back then, but we wouldn't be visiting it this time.

The bus stopped in the centre of Siem Reap. There was a parade of colourful tuk-tuks waiting, to take us to our hotels at no charge. What a thoughtful service to provide! Of course, on arrival at the hotel in our free tuk-tuk we were surprised to discover that there was now an obligation to use this driver, now charging, for all our transportation requirements around Siem Reap and Angkor Wat. The poor man was quite disappointed that he had landed the only tourists ever who just wanted to spend one night in Siem Reap and not look around. However, we did make an agreement with him for tomorrow morning.

Our hotel was 'The 1920', an attractive little place near the middle of town. It was dinner time, but we needed some money first. I had Thai baht to get rid of, so we wanted a currency exchange rather than an ATM. What a surprise to be given a handful of US dollars! There is Cambodian money, called the riel, but it takes 2700 of them to equal NZ$1 and nobody bothers with riels, at least in the cities. Now, armed with Free World Money, we went down to the Siem Reap River looking for dinner. There had been a few good little restaurants along here back in 2002. Now there were just temporary street-food stalls, but that was good enough. Sitting among the locals in the dust, on standard low plastic stools. Good clean, healthy food. Beef, cardamom chicken, hollow vegetable, and two beers: US$9.50. The 50 cents change from $10 was 2000 riels because they don't have US coins. Back to our quiet 1920 Hotel and early to bed.

Monday, 29 April

Siem Reap to Phnom Penh

Up early for another big bus day, 315 kilometres to Phnom Penh, supposedly seven hours. A Euro breakfast in a shady courtyard next to the quiet street, then we were into yesterday's waiting tuk-tuk and off through the crazy narrow streets and passageways of Siem Reap to the Giant Ibis Bus Terminal.

Our fellow passengers seemed to be all tourists, mostly white but a few Asians too. We left sort of on time around 8.45, out of Siem Reap quickly and onto the remarkably smooth Highway 6 to Phnom Penh. The route to Phnom Penh seems to be built up all the way, a thread of miserable poverty through a flat, featureless and appallingly hot landscape. The bus pulled in for lunch at a roadside stop, somewhere, or nowhere, between Kakaoh and Santuk, and I had a really good sour beef soup.

Time to get back on the bus, everybody happy, replete and comfortable. Then the bus wouldn't start. We all trooped off again for a short wait in the shade of the outdoor restaurant. The three bus men opened the engine cover and stood around looking concerned while the driver tried to start it. There were a couple of other Giant Ibis Buses passing through, so we ended up with a small regiment of Giant Ibis staff looking worriedly at the engine. It seemed to be a terminal problem.

Eventually, we were told that another bus would be sent from Phnom Penh to rescue us. It would be at least two hours. This was a real shame because the temperature was 37 degrees and even though we were in the shade, we couldn't escape the heat. There was nothing to do but sit on the hard seats and wait. The young British were true to form and consumed lots of Coke and fries to while away the time. In the land of lemongrass, limes and rice noodles, I can't understand why anybody would be eating fries.

I spent my time reading a bad spy novel, with occasional forays out into the blast furnace to survey the varied collection of passing traffic. Four or five Giant Ibis Buses passed through, then at last our new one arrived, and we set off, three hours late. I had been looking forward to an afternoon wandering around Phnom Penh, but that was obviously off now. Still, the driver tried to

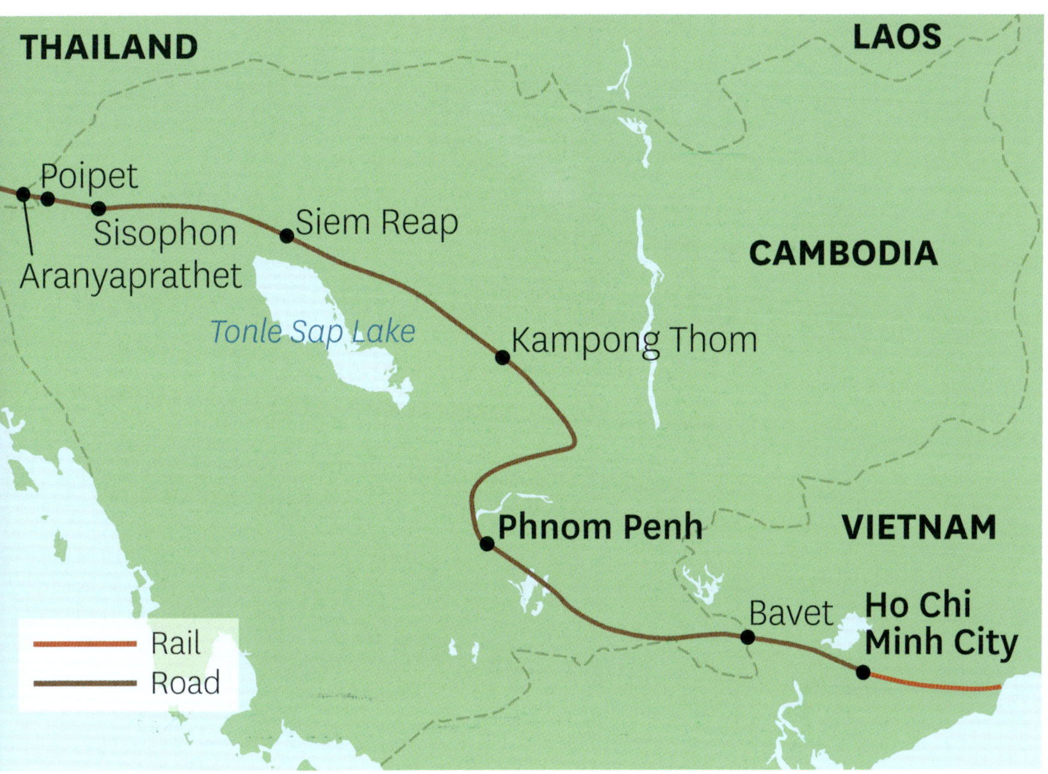

regain lost time by driving like a madman at speeds that were quite undignified for a large bus. It was real heart-in-mouth stuff, which made up for the boring monotony and heart-breaking poverty on display. In places it was like driving through a rubbish tip.

Finally, the Tonle Sap River — one of those eternal legends come to life. I've always been keen to see it. The Tonle Sap is famous for reversing its direction of flow twice a year. After the monsoon rains, the Mekong, of which the Tonle Sap is a tributary, overwhelms the flow of the Tonle Sap and forces it back upstream 120 kilometres to its source, the Tonle Sap Lake near Siem Reap. Now we followed the river for a little while, into Phnom Penh, through brand-new northern suburbs, freshly built and still being built. Endless avenues of identical empty three- or four-storey apartment blocks seemed to stretch into infinity. They looked clean and relatively upmarket but soulless.

Phnom Penh is surprisingly big, and it took us quite a while to thread our way into the centre, where it was certainly not clean and upmarket. At the Giant Ibis terminal there was a much-appreciated refund of the bus fares for everyone, US$15 each. That was dinner looked after. We took a tuk-tuk to our hotel, the Foreign Correspondents' Club. Our driver Tim was a very pleasant, happy man with a beautiful spotlessly virginal-white Bajaj tuk-tuk with car-type panelling and even little half-doors to open for getting in and

Passing traffic on Highway 6 between Kakaoh and Santuk, Cambodia.

out. It still sounded like a metal bucket full of stones though.

The Foreign Correspondents' Club was like a fantasy come true. It used to be the real Foreign Correspondents' Club during the monumental and awful events of the 1970s. Situated on a chaotic street corner across the boulevard from the wide Tonle Sap River, the three-storey-plus-penthouse cream-coloured French colonial pile is now a bar with just five or six mid-range rooms available. We had a glorious big upstairs room with old wooden flooring. The view from the sheltered deep balcony across the boulevard and the river was just magnificent, although it was dark by the time we got in.

We were pretty hungry by now; we hadn't eaten since the breakdown episode and, even then, we hadn't topped up on fries like the Brits did. So we struck out on foot in the dark warmth to a restaurant called Romdeng, about a kilometre away, the sort of place most locals could not aspire to, presumably. We sat outside in a lovely courtyard for a modern take on Khmer food. It was exquisite. The dessert was caramelized pineapple with chilli-lime sorbet, and I think it was the best dessert I've ever eaten. We walked home through what seemed like leafy suburbs in the dark. Now that it was late, the garbage scavengers were out. Whole families, parents and children were quietly going through people's rubbish. For us, this was an uncomfortable confrontation with the real world.

Straight home, a bit of time marvelling at everything from our balcony, then sleep.

Tuesday, 30 April

Phnom Penh to Ho Chi Minh City

The sun woke us at 6 am, flooding across the Tonle Sap and straight into our room. Breakfast was upstairs in the penthouse bar, semi-outside. This was absolute magic. There was the river, the boulevard, and an avenue leading away from the river full of five- or six-storey French colonial buildings in various states of dilapidation and pastel colours. It seems we were the only people in the house last night.

The bar was full of evocative black-and-white photos of the military and political events of 1975, leading up to the invasion and occupation of Phnom Penh by the Khmer Rouge, the collapse of Premier Lon Nol's America-friendly government, and the Khmer Rouge's subsequent total evacuation of the capital's population at gunpoint. During all of that the Foreign Correspondents' Club was exactly what the name says, and all the photos were taken by the then clientele. There were photos taken in the street just outside the hotel, of trucks full of Khmer Rouge soldiers driving into town past crowds of onlookers who didn't know what they were in for. It all seemed very immediate. Fascinating. And chilling.

After breakfast we had time for a quick walk along the riverfront boulevard. Wide open spaces, quite a few people, lots of pigeons and the ever-present rubbish. Looking at people, it was obvious that you're best off if you've got a military-looking uniform to wear. On our previous visit to Cambodia, Siem Reap in 2002, there was a shocking number of people with assorted combinations of missing limbs and hands and feet: the human scrap-heap caused by unexploded mines in the aftermath of three decades of war. I'm sure none of those mines was made in Cambodia. Those people were less evident in 2019; perhaps most of the mines had now been cleared. Perhaps the victims of seventeen years ago simply got to the end of their reduced life-expectancy.

On our walk along the river we got to the beautiful Royal Palace, partly

The Royal Palace, Phnom Penh.

shrouded in blue netting during a refurbishment. How a spectacular Royal Palace fits in with all the filth and poverty beats me; it's a real throwback to pre-twentieth century mores. I don't know of any other country with a mutually supportive royal family and (ex) Communist government. Cambodian politics is weird.

Anyway, back to the Foreign Correspondents' Club where Tim picked us up in his white Bajaj tuk-tuk and returned us to the Giant Ibis Bus Terminal for our 9.45 bus to Vietnam. On the way we even had a minor traffic accident: another tuk-tuk rear-ended us and rather spoilt the pristine appearance of the Bajaj. Poor Tim was understandably displeased. Onto the bus then, and we left ten minutes late on the 240-kilometre journey of six and a half hours to Ho Chi Minh City. We soon got locked up in serious traffic congestion. Phnom Penh has a real big-city feel to it, unlike anything else in Cambodia.

The trip was uneventful and a bit depressing. A major highlight was a massive new suspension bridge over the Mekong. We got to the Cambodian border town of Bavet, which was just a ripped-up dusty mess. Presumably, this is the way the Vietnamese army came when they invaded Cambodia on Christmas Day 1978 and overthrew the Khmer Rouge, doing us all a favour but somehow managing to earn the opprobrium of ASEAN, China and the West together. It certainly looks like a war zone. Of course, New Zealand, tied into increasingly uncomfortable alliances, diplomatically backed the Khmer Rouge murderers. Vietnam was the bigger enemy.

Fortunately for us, exit from Cambodia in a bus was quick and simple, just a stamp in the passport from a bored, uninterested and silently unpleasant official. Nice uniform though. Three days in Cambodia were quite enough for me.

Entering Vietnam, the bus staff took all the passports for processing and we just walked our luggage into the country, with no attention from customs. After Cambodia, Vietnam was beautiful and relatively rubbish-free. Within a few kilometres of the border, the rice paddies were immediately wet and green with growing rice.

Soon enough, we were in sprawling Ho Chi Minh City. It struck me that the place had grown up and settled down and caught up since our last visit in 2002. Back then it was a huge chaotic construction site, with growth and commerce and building constantly battering you, and concrete-mixing trucks everywhere! It was certainly not a modern Asian city at all, still recovering from the war a quarter of a century previously. Now in 2019 it seemed like the rebuild was finished, if that could ever be the case. There were lots of new high-rises, some of very adventurous design. No tuk-tuks at all, but millions of people on motor scooters.

Inspired by Alden Pyle, Graham Greene's quiet American, we checked into the Continental Hotel Saigon, right across the square from the 120-year-old Beaux-Arts style French Opera house. There was no time to explore the faded glory of the 1880 Continental right now. We set out to find some dinner and ended up at a barbecue-at-the-table place outside on the street and ordered far too much. Walking home, we watched a big fireworks display for the eve of 1 May, the great May Day workers' holiday in the communist world. The fireworks were spectacular, but the crowds were overwhelming. Solid humanity on the footpath, and solid scooters on the roads. All seemingly going somewhere. Where? Everyone seemed to be under 30. These people are really thriving. Is this what Lyndon Johnson was afraid of in the 1960s? Is this what 37 New Zealand soldiers were trying to prevent when they died in action?

May Day, Wednesday, 1 May

Ho Chi Minh City

The first thing for today was to find a good Vietnamese coffee. Cà phê as they call it. The Vietnamese have a really big coffee-growing industry; they are actually the second-largest coffee exporter in the world. They have an individual way of preparing it which they developed from the French drip-filter method. So we found a very trendy little specialist café nearby with all sorts of blends written up on the blackboard. I had a cà phê sữa đá: a little glass is filled with ice and a bit of sweetened condensed milk, and the coffee basket is put on top and filled with hot water. The water slowly makes its way down through the coffee in the basket and plops in individual drops into the condensed milk. Much patience is required, but the result is like an exquisite iced Kahlúa and milk, without the alcohol. So I went traditional, but they also had every kind of cold-press and filter system you could imagine, as well as espresso. These guys were fanatics.

After coffee we needed breakfast. Breakfast on the street in Vietnam is always special. Last time we were here there were little tables and chairs on street corners everywhere with a few contented phở-slurpers served by an invariably graceful older woman. You could get her to add a bleeding chunk of pig's liver to the broth if you wished. I remember old men offering me nips of their breakfast vodka . . . Well, that all seems to have disappeared sixteen years later. We couldn't find any food stalls in the streets around the Continental. Have they all been shooed away? We eventually settled for a very genteel looking café, all wood panels and windows onto the street at ground level, with elderly men sitting at their habitual tables reading the newspaper. I ordered phở of course and it must have been the best phở ever: beautiful fresh rice noodles, lovely rich stock, fabulous herbs. No pig's liver though.

Time to hit the streets. Ho Chi Minh City now seemed like just another big affluent Asian city and had lost much of its old scruffy narrow-street charm but is now maybe a better place to live. The awful heat bore down on us as we marvelled at the changes that had taken place since our previous visit, positive and negative.

Ho Chi Minh City.

We tried to get into the Opera House for a look, because it's supposed to be exquisite on the inside, but they weren't interested in us. Even our New Zealand Symphony Orchestra cards didn't help. Another Eurocentric attraction was the Notre Dame Cathedral, a Romanesque French fairytale with a couple of 60-metre towers, built 140 years ago. Standing inside looking at the beautiful stained-glass windows made in Chartres, you couldn't believe you were in Asia at all. I don't know how it survived a century of war and insurrection, but it's still a working church with a dedicated congregation. Another relic from the past was the similarly aged Post Office which we actually used for sending postcards. It featured a huge hall with a gloriously vaulted ceiling, and a big portrait of Hồ Chí Minh gazing down benignly on the postal proletariat. However, all this lovingly preserved colonial grandeur needs to be taken in the context of what a nightmare the French actually were for the Vietnamese.

Now we found ourselves sitting in a little restaurant named Lemongrass, for Vietnamese groovers and Western pensioners. The place had Wi-Fi, and while we were sitting there a long email from Matt at Monkey Business in Beijing arrived. Monkey Business is a British travel company working in Beijing and we had booked the whole train package from Beijing to Paris through them. You need an invitation to get into China, and into Russia, and it's very convenient and affordable to bundle all that stuff together with the train bookings and leave it to the professionals. With the next few days being on trains without internet, then straight into China where internet is a big problem for Google customers like us, today was pretty much the last day to get, from Monkey Business, the details of ticket collection and the like in Beijing. Now, like clockwork, here was Matt producing the goods.

We walked back to the Continental to collect train tickets, luggage and laundry. Tonight's train had been booked through the company Vietnamimpressive and now here were the tickets, delivered to our hotel with a few hours to spare. Then it was time for a taxi to the station. Ga Sài Gòn. Of course, the Vietnamese word for 'station' is derived from the French. And the city's pre-war name is still surprisingly prevalent.

The excitement around a late-night sleeper train departure! The shining red, white and blue carriages stretching in the dark into the distance; the platform puddled wet and reflective with tropical rain. We were quite early, and our four-berth cabin was still empty. The decor was clean white plastic softened by fake flowers on the window-side table. Four bunks, two up two down; we had the two convenient lower ones. I wandered off to buy water and beer, and when I came back a Vietnamese father and young son had moved in upstairs. The Unification Express, Train SE 2, left dead on time at 9.50 pm, for its 1726-kilometre journey to Hanoi, still on the 1-metre gauge that seems to be the standard everywhere in South East Asia except Indonesia. It was scheduled to take 31 and a half hours. Our roommates, after saying *xin chào* (hello), kept pretty much to themselves. As the train trundled out of Ga Sài Gòn in the dark, we opened a couple of beers, sat back and relaxed.

CHINA
Nanning
Đồng Đăng
Pingxiang
VIETNAM
Hanoi
Ninh Bình
LAOS
Gulf of Tonkin
Hainan Island
Huế
Hải Vân Pass
Đà Nẵng
THAILAND
Aranyaprathet
Siem Reap
Poipet
CAMBODIA
Santuk
Nha Trang
Phnom Penh
Bavet
Ho Chi Min City
Rail
Road

Thursday, 2 May

The Unification Express, Vietnam

A quiet day of train travel and reflection. After a sound night's sleep we woke up before 6 am at Nha Trang Station. There was minor commotion as our father-and-son team and the rest of the family from next door got up, packed and left the train. Nha Trang is a famous beach resort, but for me it was just an interruption and I turned over and went back to sleep. An hour later the breakfast trolley came by; now it was worth stirring. Black coffee, instant noodle soup and a boiled egg, sitting on our beds watching the rice paddies and mountains pass by. This was pretty much the pattern for the whole day, flopping about, reading, sleeping, looking. At some stage we got a new roommate, a quiet man who wasn't interested in us. Lunch came, chicken and rice with a sliced green vegetable. It wasn't very good.

We passed through Đà Nẵng and then hit the coast for the famous Hải Vân Pass. Hải Vân translates as Ocean Cloud, and in this scenic part of the route the railway winds tightly up and around the mountains right next to the sea. The train is forced to a slow crawl, but on a day like today when the Ocean Cloud was absent, the views were worth every minute. The train snaked around the hills in front of and behind us, and below us the south end of the Gulf of Tonkin stretched away towards invisible Hainan Island over the horizon.

The Gulf of Tonkin was the scene of the infamous fabricated naval action in 1964 that the US used as justification for going to war against North Vietnam. To the south of us we could see Đà Nẵng shining in the distance. It looked enormous, but its population is supposedly only about one and a half million. The old imperial city of Hué came next, another of those familiar names from grim news reports in the 1960s. It's well worth a visit, but we'd been here before. We were just heading straight through to Hanoi to connect up with a train to China.

The Hanoi to Saigon Railway was built by the French and opened in 1936,

On the Unification Express.

just in time for the Japanese invaders to use it as a major logistical asset in World War II. By the end of that war, the rail connection between Hanoi and Saigon had been severed and it stayed that way until the end of the Vietnam War. Then the government saw reopening of the wrecked railway as a crucial priority for reunification. The line was reopened just 20 months after the end of the Vietnam War, or the American War as they call it, on the last day of 1976. The Reunification Express has been running ever since, on a single-track 1-metre-wide thread from almost one end of the country to the other. It also happened to be, in 2019, the furthest you can go, directly and solely by rail, from London's King's Cross.

We were perfectly comfortable in our four-berth cabin. It was much roomier than the Australian trains, which was just as well as there was no lounge car or dining car, so we spent the entirety of our 33-hour journey cooped up in our cabin. There are no actual seats, just the four fixed bunks, but with an assortment of pillows and rolled up bedding you can sit very comfortably. And there's sort of room for two meal trays on the little window-side table. No room for knees under the table though; that's where all the luggage is. Our man upstairs stayed on his perch for the whole rest of the trip.

The dinner trolley appeared about six o'clock. The meal was identical to lunch; the second time round it tasted even worse. Still, it was only NZ$2.30 each. After dark there was nothing to see out the window except for your own reflection. Thank goodness for a good supply of books, because sleep didn't come easily after a day of sitting around doing nothing. It's rather comforting, though, to be tucked up in bed in the dark with only your reading lamp on, being rocked about while the world roars past invisibly outside.

Friday, 3 May

Hanoi

I was looking forward to spending a couple of days in Hanoi. It still has enough low-rise outdated charm to counteract the newly burgeoning commerce and traffic chaos, and you can really feel you've stepped back a hundred years. We were here in 2003 and the place left a lasting impression on me then. First impressions were not so good this time. We woke up in the dark for a 5.30 arrival and by the time the sun threw some reluctant murky light on the scene around 6 am, there was no sign of Hanoi anywhere. Just miserable fog and rain. We finally pulled into Ga Hà Nội one and a half hours late, around seven o'clock.

By the time we got out the front of the station, the weather had cleared a little and we decided to walk off our two-day confinement. It was perhaps a couple of kilometres to our O'Gallery Premier Hotel. We were ushered to a quiet corner and someone came and very apologetically informed us that we couldn't stay here, because our room had been flooded out in a recent shower of rain. But they would put us in their brand-new sister-hotel, the Oriental Jade, where we were checked in at 8.30 am.

The receptionist even told us where to go for the best phở in town, coincidentally just around the corner. Phở is supposedly actually a Hanoi invention, and we were certainly ready for breakfast, so, following instructions, off we went to a busy little place on a corner nearby. The broth was different to Ho Chi Minh City phở, more delicate and less herby. The place was packed out; we shared a small table with two well-heeled, well-dressed, well-made-up ladies who phở. They were very friendly through a curtain of broken English, and quite enthusiastic to see us enjoying their cuisine. They shared their quẩy fried breadsticks with us then disappeared, perched incongruously in their finery on a shiny little scooter. Time for a good flat white in a nearby café, then a walk around the lovely Hoàn Kiếm Lake.

Hoàn Kiếm Lake is a picturesque body of water in the centre of the city; to

Hanoi. Funeral procession of the former President, Lê Đức Anh.

walk around it is perhaps a couple of kilometres. As you walk, the view of the lake, the trees and the surrounding French colonial architecture constantly changes, and there is the haunted-looking Turtle Tower on its own little inaccessible island. Hoàn Kiếm means 'The Returned Sword'. The relevant legend involves a large turtle, but apart from that there are remarkable parallels with the sagas of the Volsungs and King Arthur.

Suddenly there were policemen everywhere clearing the road next to the shore of the lake and putting up green crowd-control ropes. A few dark-coloured SUVs with tinted windows appeared on the road, then some open military jeeps, and a column of big heavy Soviet-made army trucks, looking like something out of the 1970s, slowly roared past, belching blue smoke. The trucks were immaculate glossy olive green with massive whitewall tyres. Their cargo was squads of soldiers, sailors and airmen in dress uniform with peaked caps, carrying weapons. The police on the roadside all stood to attention and saluted. After the troop carriers came similar trucks covered with garlands of flowers, and I began to see where this was going.

Sure enough, the last of the trucks carrying flowers was towing a gun carriage with a coffin in it. This was the funeral procession of the former President of Vietnam, General Lê Đức Anh. Lê had also been the commanding general during the invasion of Cambodia in 1978. After the coffin came trucks loaded with assorted bric-a-brac which I thought might be the President's possessions to accompany him in the afterlife, which isn't a very communist concept. These trucks were followed by buses which looked like a battalion of motorized mourners. When the column disappeared in the distance, everyone started moving again and all that was left of the President was a cloud of foul-smelling blue haze.

We continued through the afternoon exploring the extraordinary chaos that is central Hanoi. We managed a visit to the big statue of Lenin: Lé-Nin. He was surrounded by kids playing football, and because of the pose the sculptor struck, appeared to be joining in. It was a very pleasant afternoon, despite the maniacal traffic, mainly because, for the first time in Asia, the temperature was manageable.

My plan for dinner was to eat chả cá. This is basically fish fried with turmeric and mountains of dill and seems to have been originally available only at the restaurant Chả cá Lã Vọng. Now there are plenty of inferior rip-offs claiming to be the real thing. It's impossible to get into Chả cá Lã Vọng without a long-standing reservation, but the management has opened a new branch nearby, Chả cá Thăng Long, and we got in. The food was definitely worth all the fuss. Chả cá is the only item on the menu; if you don't like fish or turmeric or dill, you'd better go somewhere else. They don't even cook it; you have to do that yourself at the table. It was the perfect end to a day in Hanoi.

Saturday, 4 May

Hanoi and Ninh Bình

Our fifth and final day in Vietnam. Today we were going on a guided tour to Ninh Bình, 100 kilometres south of Hanoi. This was offered by the agency through which I had booked tonight's train to China, and it seemed a good idea as we had to check out of the Oriental Jade in the morning and the train wouldn't be leaving for China until 9.20 tonight.

So our guide Bruce Lee picked us up early in a comfortable van; we collected a Korean couple with two gorgeous little boys from another hotel, and off we went. It was very foggy, warm and drizzling lightly. It took forever to get through Hanoi's monstrous traffic, then eventually onto the motorway and after that it was plain sailing. Hanoi is rapidly developing a comprehensive motorway system on a huge scale. When we were here sixteen years previously there was only one motorway, to the airport. There weren't many cars on it back then but actual bullock-drawn carts. On a motorway! Now it's all very different: lots of cars, and there's nowhere for the bullock-carts any more.

After a couple of hours being entertained by the Korean kiddies, lumpy sandstone mountains began to loom out of the mist and we hit charming but scruffy Ninh Bình. Here we met the proverbial 'Local Family' in their spacious light-filled house in a semi-rural area. They gave us a cup of tea and we set off on bicycles to Bích Động Pagoda, a half-hour-or-so ride on fairly deserted roads. This was all a wet area of lakes and streams, and the Buddhist pagoda complex itself, built in 1428, was in the mountains with lily ponds at the bottom and shrines in grottos and caves by the track up the hill.

Back on the bikes, back to the Local Family, for a wonderful home-cooked lunch in the garden patio with the Koreans; we were gradually getting to know them through a language brick wall. Then we said grateful goodbyes to the Vietnamese Local Family, just like yesterday's guests did and tomorrow's will, and we were back in the van for a while to Tam Cốc for a rowing boat ride on the inland waterways. This seems to be something very popular, with big crowds of Vietnamese lining up for the little two- or four-seater boats. They are rowed forwards by poor women who really work hard. The scenery was stunning: cliffs, mountains, gorges and water, just like a seventeenth-century Chinese silk ink-wash. Very beautiful, and rather commercially exploited, but we were on a guided tour after all. The movie *Kong: Skull Island* was filmed here in 2017.

Following this was the long van trip back to Hanoi. We had left our luggage at the Oriental Jade, so Bruce Lee dropped us there in the evening and amazingly they gave us a room to shower and get changed in, free! This was all to compensate for the supposed inconvenience of being rained out of their sister O'Gallery Hotel yesterday morning, and it was very generous. Freshly cleaned up, we had dinner at a fifth-floor balcony table nearby, then into a taxi to Gia Lâm Station, for the overnight train to Nanning in China.

Ga Gia Lâm is over the other side of the Red River and nowhere near Ga Hà Nội, Hanoi Station. Ga Gia Lâm is the only station in Vietnam for trains from China. This is because the Chinese use the Standard Gauge of 1.435 metres, unlike the Vietnamese with their 1-metre railway. The Chinese Railways have one spur of 1.435-metre gauge through the Vietnamese border and down to Hanoi at Gia Lâm. Incidentally, this is the way Kim Jong Un came to Hanoi for his ground-breaking but ultimately irrelevant meeting with Donald Trump three months previously. The taxi ride to Gia Lâm took much longer than I expected; the driver asked for more money than we actually had in đồng. So I gave him my whole pile of đồng, and he laughed and accepted it. Vietnamese đồng must be just about the most devalued currency in the world; NZ$61 will get you 1 million đồng.

Ga Gia Lâm was a rather small one-storey stone building with seemingly only one platform: the night was very dark so it's hard to be sure. It was a bedlam of hoicking, spitting, smoking, shouting, poor-looking people. There was no train to be seen anywhere. I realized that our ticket, filled in by hand, was completely illegible to us. There was no way of knowing which carriage or cabin we should be in. And no one who spoke any English.

Finally, the dark green Chinese Railways train, which the Vietnamese call Train MR 1 and the Chinese call Train T 8702, crept in through the dark under hollow lamplight, and somehow with a little help we made it through the ruckus of people onto the train and into the right cabin. The first task was to make up the two bottom bunks with the clean sheets provided, then find space for the packs, always a problem in a four-berth cabin. Next, work out

Ninh Bình. Anne observing somebody's dinner preparation.

where the toilets are — squatting only — then climb into bed. The two top bunks were occupied by an uncommunicative Chinese girl and an equally silent Indian or maybe Iranian man. The train left Gia Lâm two minutes early at 9.18 pm, and headed north into the night, for Nanning, only 400 kilometres but eleven and a half hours away.

Sunday, 5 May

Đồng Đăng to Hong Kong

Sleep didn't last long. We were rudely awakened soon after midnight at Đồng Đăng on the Vietnamese side of the Chinese border. Vietnamese border guards woke everyone up and made us pack completely then forced us all, half asleep and staggering with all luggage, off the train and into the emigration hall where there were already two very long queues. Fortunately, Anne got out of the train in a hurry and was near the head of the queue so I joined her there. Things moved very slowly. There were only two unhurried and rather menacing passport stampers for the whole train, and a few military guards to keep everyone in order. One particular guard controlled the queue in the most unpleasant silent manner. He used his long baton to indicate his instructions. Move forward. Not that far. Stop there. Go that way. He looked like he would have been happier with a stock whip.

When we made it to the passport stamper, he took his time over us, separately, and looked almost sorry that he couldn't find anything wrong with our passports, even after thumbing through and reading each page. We got through in about half an hour. I was still half asleep and had a little headache, perhaps from dehydration, and climbed onto the train and collapsed back into my bunk and fell asleep immediately. It must have taken another hour to clear the queue and get going again.

A short time later we crossed into China and pulled into Pingxiang, the Chinese border post. The clocks went forward an hour to Chinese time, making it 3 am. Everyone was woken again for a repeat performance at Chinese immigration. Now we were all experienced and there was a rush to get off the train and be at the head of the queue. Not so fast! The Chinese do things differently. They were processing people carriage by carriage and we were made to wait, crammed in the narrow corridor of the sleeping carriage, trapped by all the luggage, until it was our turn. I was busting to go to the toilet. The train toilets were locked, and I hopped uncomfortably from one

leg to the other for over half an hour until we were finally allowed off. My quick dash to the loo ensured that I was near the back of the queue when I came out. Anne was much further ahead but this time it didn't seem wise to push in and join her.

After some time, a busybody passenger told me that this queue was for Chinese only. I should be in the other queue for foreigners, who of course were presumably Vietnamese, though I couldn't tell the difference. So I went to the back of that queue, and I was the last person through immigration for the whole train. I ended up being stamped in by the officer for the Chinese queue anyway, because he'd run out of Chinese people. Still . . . we'd made it into China! The Chinese immigration officials were efficient and non-threatening to me, but they seemed to enjoy shouting at and bullying recalcitrant Chinese. Back to the bunks, back to sleep.

We woke up at about 8 am, opened the lovely white plastic lace curtains decorated with delicate flowers, and it was a miserable wet day outside. We went through a lot of wild forested areas, as well as rice paddies. It was all very green. The diesel locomotive, only a couple of carriages ahead of us, was incessantly tooting its horn, just like trains in the rest of Asia. Most Chinese trains are now non-smoking, but on this one the carriage vestibules were designated smoking areas, so there was foul cigarette smoke all the time leaking into the cabins. We arrived in Nanning exactly on time at 10.07 in the morning, which was not at all what I had been led to expect, by various online moaners. There was a three-hour layover here before our next train, to Guangzhou.

We had an important task to complete here on arrival at Nanning: I had booked and paid for the next five days' train travel, through the agency China DIY, from home. They had sent receipts with QR codes, barcodes, passport numbers . . . and now here at Nanning Station I had to redeem them all for actual train tickets. We found the ticket office and I hopefully handed over receipts and passports. Faces and receipt details were matched with passports, then scanned into a machine which spat out eight train tickets which would get us from here to Xi'an over the next five days. Brilliant service from China DIY, *fēicháng gǎnxiè!*

After a good sit-down lunch, we needed to find out where to go in this huge station to find our train to Guangzhou. Deciphering the information boards was a test first time round, but we solved that one and found ourselves upstairs in a large waiting hall, absolutely jam-packed with other people wanting to go to Guangzhou. The standard Chinese practice for high-speed trains seems to be that boarding starts eighteen minutes before departure and there's a rush down the escalators to the platform and onto the train.

Today's route across the border to Hong Kong had recently been greatly simplified: Nanning to Guangzhou, and then the brand-new Guangzhou–Hong Kong high-speed railway would whisk us in silent comfort straight to the new and fabulous Hong Kong Kowloon West Station. This was still being finished in late 2018; I followed it with interest. It caused great protest

Guangzhou South Station. Out into the fresh air to board the train to Hong Kong.

in Hong Kong, as many Hong Kongers were fearful of the introduction of a seamless land route between Guangzhou and Hong Kong. They saw the location of Chinese immigration officials on Hong Kong territory inside the Kowloon West Station, operating under Chinese law, as another nail in the coffin of Hong Kong's notional independence. As we now know, they were dead right.

I hadn't even heard of Nanning before I started planning, but it's one of those medium-large Chinese cities with a population of over seven million people, and it turned out to be a crucial junction on the route from Wellington to Alaejos. We descended the escalator to our futuristic-looking duck-snouted high-speed train, D3625. It was very long, basically two separate full-length trains joined together with two power units snout to snout in the middle, as well as one at each end. This was not immediately obvious to us. We got into the wrong carriage and told some people they were sitting in our reserved seats. They nicely pointed out that we were at the wrong end of the double train, and we had about another half a kilometre to walk to our carriage. Talk about feeling stupid.

We had to walk quite fast to get there on time but were ensconced in comfortable first-class seats when the train glided silently out of Nanning absolutely on time. It would be 576 kilometres to Guangzhou South, exactly four hours. We cruised at a leisurely 205 kph through beautiful rice paddies and mountains, although there was also a lot of fog. We came into Guangzhou over some unbelievably high viaducts, which felt like flying, and there were other high viaducts way below us.

Guangzhou had the same lousy weather today as Nanning, but we didn't go outside. We had an hour and three quarters before the connection to Hong Kong departed. Guangzhou South Station is enormous, but we knew what to do this time and we already had the tickets, so it was all very relaxed.

Soon we were heading down the escalators, outside into the dusk and onto our brand-new G train, G6545 to Hong Kong, 140 kilometres, and even faster than the previous D train at 305 kph. There were a few stops, and just about everyone got off at Shenzhen. We arrived at Hong Kong West Kowloon at 8.10 in the evening on a virtually empty train.

The station is a complete international border in itself. Chinese emigration downstairs, then upstairs to Hong Kong immigration, which was quiet and quick. Into the main hall, and the interior of the building is at least as impressive as the publicity promised. All white beams, buttresses and glass far above, rather like a hollowed-out Sydney Opera House. We took three MTR trains, then out into a rather blustery lightly raining harbourside night at Whampoa and it was just a couple of hundred metres to the Harbour Grand Kowloon Hotel, around 9.30 pm. They'd upgraded us to a suite. Luxury! It is a rather posh place and I don't think they expected us to walk in off the street out of the rain, towing backpacks after a 24-hour train journey from Vietnam.

Getting here from Vietnam had been an epic of discomfort, uncertainty and unfamiliar connections, but the whole thing had worked perfectly and was a lot of fun. We fell into our massive bed and that was the end of the day.

Monday, 6 May

Hong Kong

The day started badly for Anne; she felt quite unwell. She developed a bit of a cold yesterday, and now it hit her with full force. It was obvious that she would be bedridden for the day, and it couldn't have happened at a better place. Apart from feeling miserable, she had the luxury of a day in bed looking at that stunning view: across the busy harbour to Hong Kong Island, and all the way around to the extraordinary tongue pointing into the sea that used to be Kai Tak Airport. We never saw Victoria Peak though, as it was lost in the low cloud. I went out to get some takeaway breakfast for us. There was a good-enough Euro-style cafe a five-minute walk away and I ordered some pastries. Then I discovered I had left the Hong Kong dollars in the room, so back to the hotel, up to the eleventh floor to get the money, back down again, back to the café, get the breakfast, back up to the room; it was a well-earned breakfast by the time we had it.

We had booked a Cantonese dumpling-making lesson for the two of us. Anne was in no state for it, so we decided she would stay in bed and I would venture out on my own. Off I went on the MTR to Shau Kei Wan, at the eastern edge of the MTR system on Hong Kong Island. This was where I would meet my dumpling teacher Felicity. Being overly eager, I got there one and a half hours early, but that gave me a chance to have a good look at this fascinating residential harbour area, a sort of outer suburbia. It was very interesting from a maritime point of view, with lots of raggedy little wooden boats plying back and forth. There were even signs for ship-building companies, though I didn't see any actual ship-building taking place.

The backdrop to all this was plain residential high-rises crowding the harbour, and more being built, and behind that green hills looming up into the mist. I found a little Buddhist temple, Tam Kung, and had a look around in there, then it was time to get back to the MTR station and meet Felicity. Outside the station was a pleasant shaded little park area with trees and seats,

and elderly men, mostly, sitting passing the time. I enjoyed sitting there for a while too and observing. It looked like a genteel and gentle lifestyle. Into the station then, to the carefully described rendezvous point.

Felicity was thirtyish, sweatshirt and jeans, and a pretty enthusiastic food person. I was the only student in the class today. Our first task was to go shopping for ingredients. At the vege market we bought only little dark green-and-white bok choy, but Felicity made a point of showing me all sorts of different vegetables and jars of pastes and pickles. She was quite surprised that I was familiar with most of them. Next was the Koon Ming Ng Fung pork butcher where neat bits of dismembered pig, including a snout attached to a boneless face, were hanging on hooks. We were after organic Iberico pork shoulder, which the man minced on the spot. I'm sure I can't buy organic Iberico pork in Wellington and I didn't expect to find it in Hong Kong either. It must have been expensive, and it made me feel better about what we'd paid for the class.

Buying organic Iberico pork, Shau Kei Wan, Hong Kong.

The cooking took place in Felicity's sister's apartment. It was tiny. The sister shared it with their father. I met them both; the whole adventure was looking like a study of Hong Kong domestic life, and I found it fascinating. The kitchen in the apartment was too small for two people so we cooked on a small fold-out table in the miniature lounge. Heat came from a portable one-element electric cooker; there was no gas so no wok, just a non-stick frying pan and a pot of boiling water. We didn't make the dumpling wrappers, as we had bought them too. Fresh, of course. Yellow ones with egg, and white ones without.

We got into the wrapping. Felicity looked like she could do it in her sleep and she demonstrated all sorts of interesting shapes. I think I picked up on the basics, but I haven't practised as much as I should have since that day. The eating was a revelation. The high-quality, slightly fatty pork was a moist squirty sensation. That was my dumpling lesson then, and lunch; it was time to go. Felicity walked me back to the MTR station. I timidly mentioned Hong Kong politics; this was before the political storm that hit Hong Kong a month later. She said she knew nothing about any of that. She was an intelligent woman; I didn't believe her. But I left it at that.

Back on the MTR, my plan was to take the West Rail Line to Tuen Mun, about 50 kilometres away in the New Territories in Hong Kong's north-west. This trip features in the Lonely Planet *Amazing Train Journeys* book and

I thought it was a must-do since I was here. So off I went, with changes at Central and Nam Cheong. Well, it was all a bit of a bore actually, and I think the MTR to Hong Kong Airport on Lantau Island is more interesting. At the end of the line at Tuen Mun I didn't even bother to get off. Tuen Mun, incidentally, was close to the scene of some particularly nasty violence between democracy protesters and Chinese Communist Party supporters just a few months later. So I stayed on the train, a few changes back to Whampoa, the Harbour Grand Kowloon, and Anne. Anne was still feeling rough, but dinner in a little place nearby picked her up a bit. Good Cantonese comfort food and we both felt better for it.

Tuesday, 7 May

Hong Kong to Hangzhou

A big train travel day today. From Hong Kong direct to Hangzhou, 1800 kilometres to the north-east, in seven and a half hours, which implies an average speed of about 240 kph. It was another grey and blustery day on the Fragrant Harbour, the English translation of Hong Kong.

Anne and I had been in China several times before. In fact, my first visit was with the New Zealand National Youth Orchestra in 1975, while the Cultural Revolution was still happening. We left the British colony of Hong Kong this way back then too, but the experience was quite different. We all took a slow suburban train to the end of the line at the tense Chinese border at Lo Wu, where unsmiling People's Liberation Army soldiers in baggy khaki, toting rifles, escorted us on foot over the Sam Chun River Bridge to China. A huge train (I'd never seen 1.435-metre Standard Gauge before!) was waiting at a village at the other end of the bridge, towering over us.

That place has since become the monstrously built-up megalopolis of Shenzhen, one of China's main commercial hubs. There were just a few houses in the village then, and one government building where we were given lunch. It was the first time I'd ever used chopsticks: it was either adapt or starve. After lunch (still in 1975), we boarded our comfortable private train to Guangzhou, which we called Kwang Chow back then — and before that it was Canton. On the train we were served bottomless tea in big cups with lids, by petite immaculate stewardesses.

After concerts in Kwang Chow the orchestra flew to Peking in a brand-new Boeing 707 that smelt like the inside of a new car. The stewardess serving ice-cream proudly told me that the pilot earned the same amount as a greengrocer. And then he took three attempts to land at Peking. Perhaps he should have stuck with selling bok choy. That tour was a major New Zealand–Chinese diplomatic event. One of the soon-to-be-vilified Gang of Four, a man who had responsibility for the revolutionary art and culture of the proletariat,

G-Train departing Hangzhou East Station.

came to our Peking concert. I shook his hand. That was Zhang Chunqiao.

Fast forward to the twenty-first century. The attraction of Hangzhou was the famous West Lake. It's an old human-made expanse of fresh water, about 6 square kilometres, on the edge of the city of Hangzhou, with beautiful pagodas, artfully placed willows, reflections, moonlight on the water. The lake is renowned throughout China for its beauty, and this reputation rubbed off on Hangzhou. In the thirteenth century Marco Polo described Hangzhou as 'without a doubt the finest and most splendid city in the world'. Nowadays, Hangzhou is just another sprawling, ugly, polluting Chinese metropolis of 10 million people, but it still has the lovely shallow lake, largely unspoiled, dominating its western boundary. And West Lake Fish, a culinary delicacy harvested from the lake — certainly worth getting off a train for.

Now, on train G100 to Shanghai via Hangzhou, we whispered out of Hong Kong, underground, beneath Lo Wu and the old Sam Chun River border bridge. We didn't see the bridge, but I believe it remains there, preserved as an historic site. We headed back the way we had come two days ago. From Guangzhou we went directly north to Changsha, then turned north-east for the long run through Nanchang to Hangzhou. Despite the blistering speed, the trip was mundane. The air was very misty all the way. Some of the route was rural, but big cities were dotted just everywhere. Gaunt, soulless high-rises were plonked onto the ground in packs of 20 or a few hundred, and no wilderness or forests, just housing for people, or growing food for people. This is of course one of the oldest civilized places on earth. People have been living here constantly, much like this, for five or six thousand years.

At 6.25 pm, we pulled into Hangzhou East Station, exactly on time. We needed a taxi to our hotel, and I was well prepared with the name and address of the Midtown Shangri-La Hotel in Chinese characters on my phone. Some Chinese taxi drivers can be awfully rude and uncooperative. We got to the head of the taxi queue, a taxi pulled up and we put our luggage in the boot, climbed in and I showed the driver the screen of my phone. He started shouting angrily. Either he couldn't read or he didn't want to go there. There was no assistance from the uniformed men outside, so we got out and removed our luggage and tried the next taxi. Fortunately, this driver just nodded cursorily at my phone, so he was our man. It was quite a long way to the Midtown Shangri-La.

We ventured out to find dinner. This part of town was quiet, but we found a busy little strip of humming restaurants. We walked into one and a waiter showed us to a table. But before we could sit down, someone else sat there. Our waiter immediately lost interest in us, and we couldn't get anyone else's attention, so we just walked out. We took a big breath and walked into the next place. This was much better. We were taken to a table and the English Specialist Waitress was summoned. She hardly spoke any English, but she had an English translation app on her phone. She was lovely and very determined that we should understand the entire menu, which was only in Chinese with very few pictures.

It turned out that this was a hotpot restaurant: a big steaming pan of stock was brought to simmer on an electric element recessed into the middle of your table and you dipped the various meats and veges that you had ordered into it to cook. It's a very popular and sociable Chinese eating habit, similar to fondue but definitely without the cheese. Our waitress slaved over her phone, but the app was only partially helpful. Somehow 'chicken stock or fish stock?' became 'chicken bottom soup or fish bottom soup' and it took me a while to work that out. I didn't really want chicken bottom or even fish bottom. There was lots of confusion about veges too, but with help from other tables, we ended up with a massive basin with beef stock (cow bottom soup) with lots of beef, potato slices, lotus slices, heavy flour dumplings and a mountain of coriander. And two beers. It was magnificent. Quite spicy. The waitress's patience and good humour had restored my dented faith in Chinese hospitality, and we staggered home in the dark, fortunately not too far.

Wednesday, 8 May

West Lake

West Lake was just a ten-minute walk away from the Midtown Shangri-La, mostly through parkland. We hadn't seen it last night; when we got to the embankment now, we could see what all the fuss was about. A vast expanse of unruffled water under a leaden grey sky, leading the eye to a backdrop of hills disturbed by the odd pagoda. Nearby, small groups of elderly people sat outside under the trees at teahouses, quietly enjoying each other's company. At this cool early hour, we could already see a lot of people walking on the embankments, causeways and bridges in the distance. We set out around the lake slowly in an anticlockwise direction, taking it all in. I think the full distance is about 15 kilometres.

There were the usual tai chi groups, mostly elderly: an interesting combination of old-style Chinese pyjamas, high-collared with braid and toggles on the front; and tracksuits and sneakers. Little boats for hire were setting out, just the thing for a small Chinese family outing without too much exercise. I wasn't tempted, as I needed some walking exercise after spending yesterday cooped up in a train. We climbed over the first arched stone bridge, which looked quite new and was probably a replica of something that had stood for centuries. Near the bridge men were flying beautiful paper-and-wood bird-kites. Some corners of the lake were covered in waterlilies. The long causeways were a pleasant stroll out into the middle of the lake with water close by on both sides, lined with willow, plum and peach trees.

This must all be heavenly for an inland Chinese city-dweller living in a cramped apartment in a crowded ugly place. There are famous spots which Chinese have on their bucket lists, such as 'Sunrise Over the Lake in Spring' and 'Three Pools Mirroring the Moon' which are perhaps the result of centuries-old marketing hype.

Soon it was time for a lunch of dumplings in a paper bowl, sitting on a well-placed bench contemplating distant pagodas. A man sitting on the same

West Lake.

bench started talking to us; amazingly enough he had studied at Auckland University for a while. The three of us enjoyed the opportunity to chat. Despite his foreign university education, he was now working on the lake as a tour guide, for locals. He couldn't understand why we would want to walk all the way around the lake. It seems that's not the Chinese way.

On our to-do list was the nearby famous and crowded Leifeng Pagoda, crowning a hill on the south shore. The original was built in the tenth century but fell down in the 1920s, and the new one was built in 2002. In the basement the foundations of the original are visible under glass. From the lookout five floors above there's a comprehensive view of West Lake with Hangzhou menacing behind, but the lake is not really suited to the elevated grand view. It's better from water-level, appreciating both the micro and the long view, and then its attractions are infinite. By now the crowds were really growing as the day warmed up. There was a sign with a moveable pointer, like a New Zealand bushfire warning sign: Tourist Density: Moderate!

As we wandered up the city side of the lake the communal activity became extraordinary. There was a large choir with a conductor and a small ensemble of accompanying traditional instruments, very enthusiastically singing Chinese pentatonic music in unison. They had gathered a big and appreciative audience who helped out unhelpfully. Especially entertaining was the old man who showed his artistic sensitivity by conducting along with it, completely at variance with the actual conductor. As a musician, it was impossible to watch him. Further on, there was the sound of more traditional music, recorded and amplified this time, with three camp old men dancing with colourful scarves, hats, flowers and outrageous sunglasses. They were joined by a plain old dumpling of a woman with short grey hair wearing a sweatshirt and track pants. Then we came upon the pavement calligrapher. He stood upright and with a very long brush painted lots of Chinese characters on the pavement, just using water. They slowly evaporated in changing patterns in the heat of the day.

The sunset over the western hills was stunning, no doubt helped by a good dose of Hangzhou air pollution. There were crowds of people to watch it, including regiments of photographers with big cameras and tripods.

We finally made it to our chosen Grandma's Home restaurant for dinner, glowing on the water's edge with green neon. It's recommended for its delicate local Hangzhou cuisine, famous in Chinese food circles. This includes the legendary West Lake Fish which I'd been looking forward to for decades. The real West Lake Fish is carp, caught in the West Lake. We were shown to a table, then promptly forgotten. After an age I got someone's attention and we managed to order. The menu had English translations which was a novelty. Well, they had run out of West Lake Fish. What a disappointment! I'd come halfway around the world to West Lake to eat their fish and they didn't have any. And the empty plate of the two women next to us had a mocking picked-over fish skeleton on it. We had just missed out. So we settled for Beggar's Chicken cooked in lotus leaf and clay. The place was packed and humming, and not a white face in sight despite the English menu. Walking home on an ordinary Wednesday night, the lakefront was pumping with people.

Thursday, 9 May

Hangzhou to Xi'an

We were out of the Midtown Shangri-La at eight o'clock to get breakfast at stalls along the lakefront. It was a glorious morning at West Lake, with a beautiful clear blue sky for the first time since Ho Chi Minh City. Last night's massive crowds had evaporated, and there were just yesterday morning's small groups of elderly people, all playing cards this time. Perhaps Thursday is cards day in Hangzhou. Mostly old working-class men, but a few mixed groups as well. This was great people-watching. We sat on bollards to eat our takeaway breakfasts out of cardboard bowls printed to look like rice pattern: cold noodles with cucumber, Sichuan preserved vegetables, and garlic on top. There is so much variety in Chinese street food! We sat in the comfortable morning warmth and polished it all off, watching the awakening day on the water, the card players, and groups of people exercising.

One man, not particularly young, began to dance a graceful and skilful ballet on rollerblades. Nobody took much notice, but they all left plenty of space for him and he was fabulous to watch. So fabulous that the police soon turned up complete with notebooks and wagging fingers and the implication of much more. That certainly produced a crowd of onlookers. Subversive skating was stopped, order restored, and the crowd dispersed.

Now it was time for a taxi to Hangzhou East Station, and farewell to West Lake. Today we would be using the last of the swag of train tickets we had so excitedly collected at Nanning Station only four days ago. Train G1894 would whisk us from Hangzhou, far inland to the Old Imperial Capital of Xi'an, 1470 kilometres in a G-force-inducing seven hours 20 minutes. By now, getting to the right train through the crowd was a simple enough affair, and within an hour or so we were comfortably seated for the day. The train left 20 seconds late, at 11.10 am. The final impression of Hangzhou, oddly, was a huge power station cooling tower with a fast fighter jet circling it — military and industrial might all rolled in one.

Nanjing arrived out the window, and then lunchtime, a menu in Chinese for service at our seats. Google translate came to the rescue, and we each ended up with dan dan noodles; that's spicy mince and pickled veges on noodles, simple tasty food that you can't really go wrong with. China flashed by outside at 300 kph, mostly flat country, rice and other crops, lots of cities, no wild forest. The display on the carriage bulkhead told us it was 33 degrees out there.

Towards the end of the day there was an eye-catching jumble of mountains out on the left in middle distance; this got closer and became the legendary Hua Shan, one of Taoism's five sacred mountains. Soon after Hua Shan we pulled into Xi'an North Station dead on time at 6.35 pm in broad daylight: nightfall was definitely starting to get later now. This time a taxi was no problem at all, and we drove into the city on wide busy boulevards alongside the ancient city wall. Our Eastern House Boutique Hotel was very central, inside the wall and close to the Bell Tower. It was a big black building with nothing very Boutique about it at all.

First, we had to collect some important things from reception. On leaving Xi'an in a couple of days' time we would spend the following ten days in the care of a travel agency in Chengdu who would take us from Xi'an to Lhasa in Tibet, and on to the Tibetan side of Mt Everest, then later all the way back to Beijing. Violet, our very accomplished agent at Go To Tibet, GTT, in Chengdu told us that train vouchers, and the all-important Tibet Permits, would be delivered to the Eastern House for our arrival. Sure enough, they were all there in a big envelope when we checked in.

No time to hang around now though, as we were soon hot-footing it in the dark to the Muslim Quarter, for dinner on our feet. From the Eastern House, entry to the Muslim Quarter is through the magnificent fourteenth-century Bell Tower, all lit up and imposing in the dark, in a big public square on a busy intersection. On coming out the other side of the Bell Tower, we entered a madness of garish neon, noise, huge crowds and food smells. The Muslim Quarter is exactly what the name says. Xi'an's large Muslim population is crammed into a rabbit warren of narrow streets. While the Chinese government is currently frightened of the Uyghur Muslims and is detaining them en masse and trying to socially engineer them, ordinary Chinese are happy to visit the Muslim part of Xi'an and eat there.

There was an extraordinary variety of food available and on display, all of it quite different to Western perceptions of 'standard Chinese food'. People in western China eat a lot of sesame products. They have a sesame paste, zhī

The Muslim Quarter, Xi'an.

ma jiàng, rather like Lebanese tahini. So here, in a few locations, it was being manufactured. From a distance, above all the other racket, you could hear fast, loud, percussive thumping. As you got closer there were two big men, stripped to the waist, alternately beating a big block with very large wooden two-handed mallets which they raised above their heads. The block was spread with sesame seeds which they were pounding to a paste. The men were perfectly coordinated, the tempo fluctuating between fast and slow. It was a fascinatingly different example of food processing.

Meanwhile, the food: I queued for a lamb baozi, basically a lamb burger with stewed lamb and no vegetables in it. It looked delicious and tasted about half as good as it looked. I followed that up with a few other bits and pieces, and then I was full. No matter, we continued to walk around the crazy place for ages, just observing the differentness of it all.

As was becoming normal on this trip, we finally stumbled home through the night to the Eastern House, exhausted. We were asleep in minutes.

Friday, 10 May

Xi'an

The Army of Terracotta Warriors, manufactured about 2240 years ago to protect Emperor Qin Shi-Hua through the afterlife, is situated outside of Xi'an. Once we got to the site, our taxi driver delivered us straight into the arms of a guide, which suited us just fine. I didn't want to go through the place ignorant.

Peter the guide led us through the experience patiently and unobtrusively. It takes a while for the enormity of the site, the project to conserve it, and the time scale, to set in. It's sort of gently and subtly overwhelming. The main site is protected by an enormous aircraft-hangar-type structure. We could look at endless serried ranks of life-size life-like warriors, then concentrate on the features of a few individuals, one by one. All were made of kiln-baked clay. Some were in perfect condition, many were cracked or broken, especially the ones in newly excavated pits. It's still an ongoing project, and there were people working among the figures. It was an odd sight, seeing one coloured figure moving between all the monochrome immobile warriors. This is a major tourist site and of course there were a lot of people present, but we, the living, were seriously outnumbered by the warriors and the whole place had an air of reverence and awe.

In the adjoining museum there were individual warriors and artefacts in glass cases for a close-up look. There too were the stunning bronze chariots, two of them, half scale, each with four beautifully detailed bronze horses, all two-and-a-quarter millennia old and in remarkable condition. There's ancient chromium plating on the bronze blades of the warriors' weapons and the scientific jury is still out on how that might have happened. Apparently, there's chromium in the soil around Xi'an.

After a few hours we exited through the proverbial gift shop and then things got a bit tacky. Peter completely blotted his copybook by leaving us at a house to try the local peasants' noodles for lunch. We were unashamedly ripped off for a big bowl of cleaver-sliced noodles with spam in soup, at an exorbitant price. Nonetheless, the experience of the Terracotta Army was fascinating and I learned a lot. Only the soup left a bad taste.

Back in town, we had more rail tickets to collect. GTT Violet had sent us the usual vouchers, which needed to be exchanged for real tickets. We found a Chinese Railways ticket agency in a hole in the wall just along the road from the Eastern House. Passports were shown, and the voucher, and the all-important Tibet Permit which I felt must be guarded with our lives. All being in order, train tickets were passed out through the window, for the train from Xi'an to Lhasa tomorrow.

Except the tickets to Lhasa were not actually from Xi'an. The train from Shanghai to Lhasa does in fact stop at Xi'an, but Chinese Railways won't sell tickets from Xi'an to Lhasa. GTT Violet was fully aware of this and had her own way around it. She booked us tickets originating in Zhengzhou, about 500 kilometres further back east towards Shanghai. You're allowed to book a train from Zhengzhou to Lhasa, and that ticket from Zhengzhou will allow you to enter the heavily policed station at Xi'an and board the Lhasa train there. All a bit odd, and possibly a bit dodgy, but GTT Violet had previously informed me that she regularly booked this trip for foreigners, and it always worked. Now, tickets for tomorrow secured, it was time for our final Xi'an adventure: a bike ride along the city wall.

Xi'an, a city of eight and a half million, has one of the best-preserved walls in China. Built in the fourteenth century but obviously completely rebuilt since then, it's a 12-metre-high rectangle around the old city, with a perimeter of 14 kilometres. The top is very wide, spacious, smoothly paved with stones, and perfect for bikes. We walked from the railway ticket kiosk a short distance to the South Gate, up the steps to the top and hired bikes there. The bright blue sky was clouding over and the temperature was pleasantly in the high 20s. The cycling on the flat wall was lovely and relaxing, with not too many people, and a fascinating way to see the city. The wall was a wide grey boulevard with red flags and lanterns hanging all the way, very picturesque and stretching away in the long view. We were above roof-height for the older traditional buildings inside the wall, and outside there was a jungle of high-rises, mostly grey and individually unattractive, but the whole vista was fascinating, with a certain beauty. At regular intervals along the wall were large bastions with oriental roofs, just to help remind us where we were. Outside the wall there was a wide dry moat all the way around, providing a bit of green to offset the totalitarian grey and red.

On the wall, Xi'an.

Above the North Gate we looked over the moat to tomorrow's Xi'an Station. This was not Xi'an North Station where we had arrived last night, but the old Xi'an Station. Xi'an North Station is a part of the High-Speed Rail network and regularly services a certain number of foreign tourists, like us last night. Xi'an Station, however, is not on the High-Speed Rail network. It services the older slow green trains which see very few Westerners. That includes tomorrow's train to Lhasa.

Darkness was coming on and so were our appetites. We walked straight to a restaurant near the Eastern House, for a really good hotpot. No problems with ordering this time. Beef, greens and mushrooms. For once we didn't overdo it, and, in recognition of last night's sesame bashers, there was a delicious sesame dipping-sauce.

Saturday, 11 May

Xi'an to Xining

A big day today: we're going to Tibet. Not many days in one's life start with that statement. Tibet has been a place of interest to me ever since reading *Tintin in Tibet* in the 1960s, seeing Captain Haddock blowing on a long Tibetan horn when he shouldn't have. When the railway line from Nanshankou to Lhasa opened in 2006, the world's highest railway at 5072 metres, I thought, 'I'll have to do that sometime.' The Chinese are rather sensitive about Tibet and access for foreigners is limited. The only way in is with an accredited Chinese travel agency which can apply for the coveted Tibet Permit on your behalf. I discovered GTT, in Chengdu, which incidentally is not in Tibet. GTT offered a week in Tibet that could include a long bus trip to Mt Everest. A trip from Wellington to its antipodes via the highest mountain in the world! I was sold.

Now here we were checking out of the Eastern House in Xi'an. Before we could get to reception the bellboys efficiently intercepted us and, armed with a translating app, asked if we wanted a taxi, and where to.

'Yes, please,' I said, 'a taxi to Xi'an Station.'

'You mean Xi'an North Station?'

'No. Xi'an Station.'

'You mean Xi'an North Station. For tourist.'

'No. Xi'an Station for the train to Lhasa.'

'Show me ticket, please.' I produced the ticket.

'Ah. This ticket from Zhengzhou. Is wrong ticket.'

'No, it's right. It's from Zhengzhou, but we are boarding at Xi'an.'

'Ticket wrong. Do you want to go to Xi'an North Station to rebook?'

'No. I want to go to Xi'an Station to catch my train.'

'Cannot do that.'

I broke off and walked over to reception, where the woman had a smattering of English.

'Please check us out,' I said, 'and refund our security deposit. Here's the receipt for it. And please get us a taxi.'

'Show me ticket please . . . Ticket wrong. You must change it.'

Meanwhile, the receptionist had looked up our file, to see what we were meant to be doing. I think she saw a note saying we were meant to collect the Tibet Permit from reception, but there was no record that we'd done that, and she couldn't find the permit. Of course not: we already had it. No explanation from me would work, and suddenly the illegal taxi was the least of her problems. People were sent scurrying in every direction to find our Tibet Permit. In desperation I decided that the only person who could help me was GTT Violet, 600 kilometres away in Chengdu. Fortunately, I found her direct-dial number in an email. I showed it to the receptionist and somehow I persuaded her to ring it. By some miracle, Violet answered and there ensued a harrowed conversation which turned out to be the receptionist telling Violet that our Tibet Permit was missing and we had the wrong train tickets anyway.

I asked for the phone and spoke to Violet. Such a relief! I explained that we already had the Tibet Permit; it wasn't missing. I said that we fully understood the Zhengzhou ticket situation, but the staff didn't, and they wouldn't get us a taxi to Xi'an Station. Could she please fix it with the receptionist? I told her how much I loved her and passed the phone back to the receptionist. There followed another phone conversation in Mandarin in which I saw recognition and understanding slowly start to light up the receptionist's face. Success.

'So sorry! So sorry!'

'Now give me the security deposit and check us out please.'

'Give me receipt.'

'I gave it to you.'

'No, you didn't.'

'Yes, I did. You stapled it under those pages.'

'Oh, so sorry!'

The poor woman was totally frazzled by now. Finally: 'Now please get us a taxi to Xi'an Station.'

'Yes, yes.' Then the bellboy weighed in for a last attempt:

'Departure time on ticket is 5 am. Time now 9.45. Ticket expired!'

Of course it said 5 am, that's when the train left Zhengzhou, 500 kilometres away. Stage whisper: 'Just get us a bloody taxi!!!'

At last we were in a taxi to Xi'an Station with a typically rude, rough driver who shouted at me unintelligibly when it was time to get out at the station. All signs were in Chinese only. Getting through security was okay;

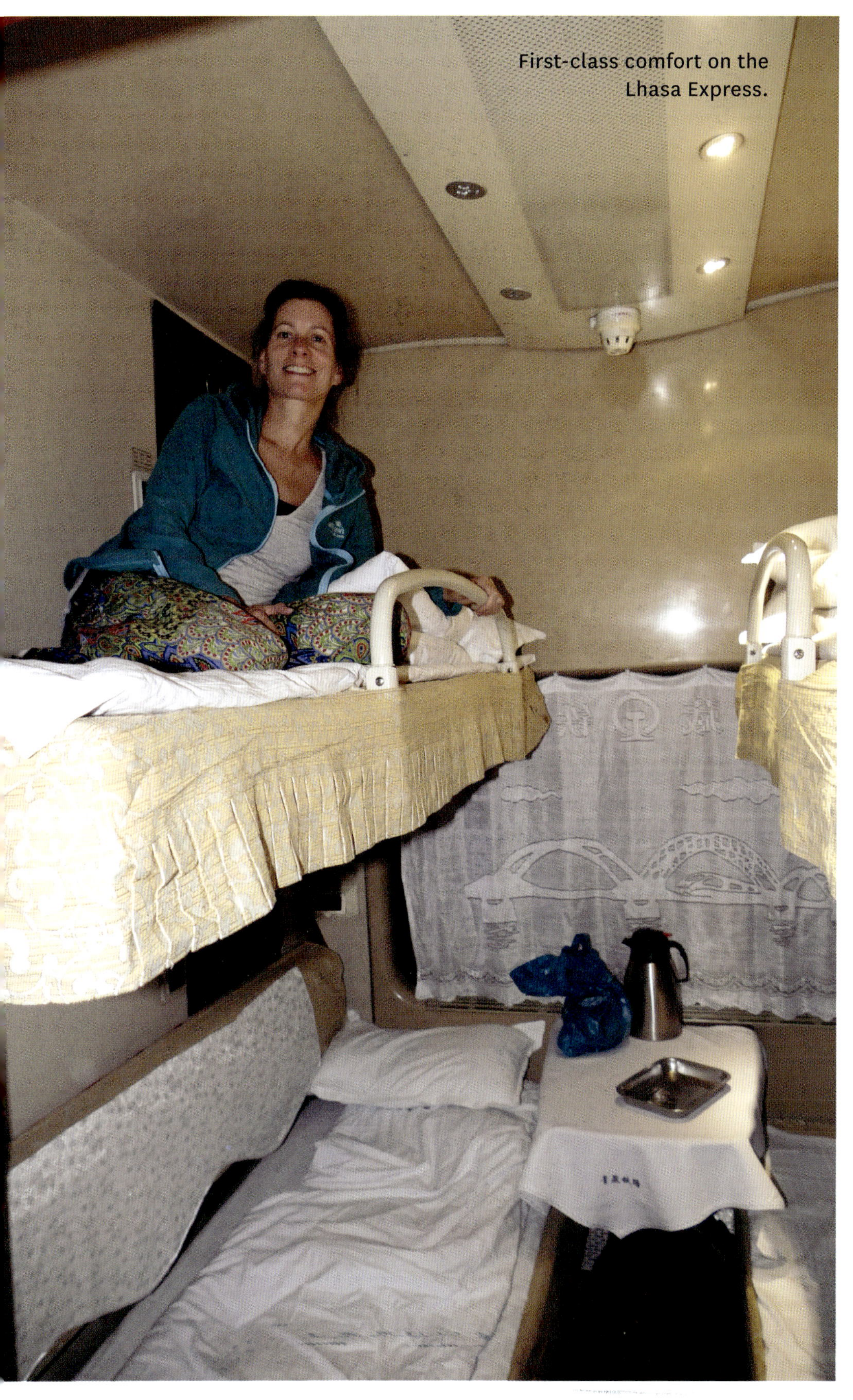

First-class comfort on the Lhasa Express.

you need a ticket to get into the station, but nobody was concerned about the Zhengzhou aspect of it. Xi'an Station was certainly older and more Chinese, less international, than what we were used to, and it was a relief to see our train number, Z165, up on the departure boards. All our trains since Guangzhou had been G trains, long-distance high-speed trains. Today's train would be a Z, denoting slower, older, long-distance sleeper.

At the carriage door there was rigorous checking of tickets, the precious Tibet Permit, and passports and visas, and then we were finally on the Shanghai–Tibet Express after a year of planning and a morning of extreme frustration. I sat down and almost wept with relief. We were in a four-berth soft-sleeper cabin, two bunks on each side, one side for us, and a Chinese couple on the other.

We departed smoothly and quietly at 11.24 am, on time as usual. This was going to be a long trip of 2864 kilometres, 32 hours. Straight out of Xi'an, heading west, the scenery was brilliant as we followed the Wei River through gorges, wild hillsides and cultivated terraces. The region was one of the ancient cradles of Chinese civilization. We weren't going particularly fast, about 60 to 120 kph and we hardly ever stopped. In the gorges it was like a model railway set, with rail lines on different levels going in different directions and through tunnels and over bridges. Once through the gorges, development was omnipresent, with industry, agriculture and railways. Railways, including unmistakable new high-speed tracks, were spreading just everywhere. The investment in infrastructure, and the speed of its progression, is just mind-boggling.

We had quite a party of five in our cabin for a while. Our roommates had a visitor and it seemed there were lots of phone messages to listen to and calls to make. Five people in a small space doesn't leave much room for manoeuvre and inevitably one of the Chinese knocked over an open hot water flask. I caught it in mid-flight, but there was still hot (fortunately not boiling) water all over them and their lower bunk. That little disaster allowed for a bit more bonding.

We were now heading north-west, approaching our first major stop for the day at Lanzhou, where we arrived about 7.30, just on sunset. Lanzhou is a crossroads city of three and a half million on the Yellow River, and supposedly right at the geographical centre of China. It's famous for a soupy dish of thick noodles and spicy beef, but we were there for only ten minutes. Soon we were underway again in the dusk. Our roommates' guest spoke a little English and he had an important message for us: 'This train does not go to Lhasa!' That came as a shock: are we on the wrong train? No, soon we would all have to get off this train and onto another one. Apparently, this was normal, and I was very grateful to our roommates for going to the trouble to make sure we knew what was going on. Sure enough, around 10 pm the train pulled into Xining, further to the north-west. Lots of announcements in Mandarin, and everybody, with luggage, piled out onto the platform.

Here in the middle of China, in the middle of the night at 1500 metres, the air was quite cool. Drawn up on the other side of the platform was another similar train and it was just a matter of crossing the platform into the same carriage number, same compartment, same bunk. Same ticket. And, naturally, another opportunity to closely scrutinize Tibet Permits, passports and visas, and tickets. Certainly, nobody has any chance of sneaking into Tibet without proper authorisation. This was all supervised by the female train staff who wore a uniform of long claret-coloured wool greatcoats, shaped at the waist. Quite a fashion statement.

Our new compartment was modern, lighter and brighter. And, I noticed, it featured the promised oxygen portals which were missing in the previous train. The high-altitude adventure begins! We still had the same roommates. As the Xining to Lhasa Express headed west out into the dark Central Asian wilderness, Anne and I each popped an acetazolamide pill against altitude sickness, climbed into our bunks and were asleep before midnight.

Sunday, 12 May

Xining to Lhasa

We woke up at the crack of dawn, about 6 am. It hadn't been a great night. The train was very smooth and quiet, but the mattresses turned out to be not in the same class. I noted that the Chinese lady in our cabin had turned on their oxygen, so I turned ours on too. It's not in a mask or anything, rather just a nozzle on the wall, similar to the fresh air vent above you in an aeroplane. A first glance out the window confirmed that we were in a new world of grandeur and immense proportions. We were on a vast plain with intermittent snow on the ground, and snow-covered mountain ranges in the middle distance. Occasional little station platforms flashed by, and their nameplates featured altitudes around 4100 metres. It was a vista of unutterable loneliness, with not a tree to be seen. As the blue sky brightened into the day, so too it clouded over, and it remained a grim steel-grey for the rest of the day. We were on the Tibetan Plateau! Heading south now, towards the Himalaya. We began to see signs of life: our first yaks.

Sometimes the highway was beside us, with lots of heavily laden trucks going to Tibet. All the trucks coming back were empty or carrying another empty truck to save fuel. That sort of summed up the strategic reality: China pumps endless resources into Tibet, but nothing much comes out. The engineering around the railway was incredible: massive embankments to keep the trains above the permafrost, huge land stabilisation works on slopes, long viaducts, bridges and tunnels. We were on a standard railway, but in places we could see a high-speed successor already taking shape.

Perhaps because of the piped oxygen, the air was extremely dry. I was drinking more water than in the tropics, and the inside of my nose felt like broken biscuits. As the day carried on, the inevitable altitude headaches made their presence felt, and a certain torpor set in, despite the acetazolamide. We filled in some time in pleasant but laborious conversation with our roommates. They were Chinese, living in a town called Nagqu. They had been back in Shanghai for a holiday, visiting relatives. They didn't like living in

Nagqu, as the air was too thin and cold. They said most Chinese felt that way about living in Tibet.

We hit the Tanggu La Pass at just over 5000 metres, then dropped down a bit and into Nagqu to say goodbye to our friends. Nagqu: what a godforsaken place! The Tibetan vistas are so huge that we first saw it from miles away, small, insignificant and alone on a vast snowy plain. As we got closer, the prospect didn't improve, and I felt sorry for those two who now called it home. They didn't look too happy either.

After Nagqu we cruised down past the big sacred Namtso Lake and eventually into the gigantic new Lhasa Station. It was 7.35 pm, and still light. Oddly, for such a huge country, China uses only one time zone. So right now, back in Hangzhou, 3000 kilometres away to the east, it was also 7.35 pm but had been dark for an hour. Out of the monstrous station building and into the crisp evening air of Lhasa. The front concourse seemed rather like a military camp, with military vehicles and squads of soldiers in full combat gear, rifles and helmets dotted around the place. No Photography! We walked over to the immigration building. China annexed Tibet in 1951, but this really felt like the full immigration experience of a new country.

There were only four foreigners to deal with: Anne, me, and two middle-aged German men. I was rather cowed by the whole threatening military business, but the two Germans took great pleasure in mocking Chinese officialdom to its face, with salutes and clicking of heels. I just wanted to get out of there while we could. *Alles in Ordnung*, we went outside and were met by a Tibetan guide who made us feel very welcome and ceremoniously put big white silky plastic scarves around our necks. Then into a big newish white van. The four of us, as the two Germans, Thomas and Rainer, were also part of our tour.

Out onto the road into town, and it was an absolute shock. Only 80 years ago Lhasa was undeveloped and almost unvisitable. Now, the extent of ongoing urban development was both awe-inspiring and just awful. Soulless standard Chinese high-rises and highways springing up just everywhere. Presumably, they were mostly for imported Chinese people, not Tibetans. Further into the centre of town, chaotic old Lhasa took over and the first sight of the legendary Potala Palace lording it over everything is literally spellbinding. After 20 minutes we were on crazy congested Beijing Road, outside our imaginatively named Yak Hotel.

The Yak was an older place in sort of Chinese-Tibetan architecture. Our quaint heavily draped dark room was perfectly comfortable and spacious. It

Lhasa Station.

was late now and we were pretty hungry, so we went straight out to dinner at the Dunya Restaurant next door. It was right next to an international hotel (ours), so of course it was touristy, but the cuisine was supposed to be Tibetan, so I was happy to give it a go. And it happened to have an English menu. Tibetan food, at least for tourists, turns out to be sort of Nepalese/North Indian mostly. I had a yak curry thali set. Thali is the Indian idea of a main bowl of curry surrounded by little bowls of accompaniments: dal, a vege, pickles, rasam, rice . . . It was very good. Yak meat could be described as having similarities with lamb and beef. No alcohol for us though, for at 3650 metres we were both completely altituded out.

Monday, 13 May

Lhasa

We met up with the rest of our ten-strong tour group and our guide for the next week. There were four other women: a Spaniard, a Peruvian, a Chilean, and a Chinese-American. Then there was John, a large conservative single man from the southern United States. Next up came our two Germans from yesterday, Thomas and Rainer: Mercedes-Benz executives living in Shenzhen. There was a young Irishman who seemed to be a professional backpacker, and Anne and me. Our guide Tsering was a charming 40-ish Tibetan man with an endearingly gentle manner. I was most drawn to Tsering, whom I found fascinating.

Now we all walked the short distance to Barkhor Square in the middle of Lhasa to see the Jokhang Temple which Tibetans believe to be the world centre of Buddhist worship. Barkhor Square was full of poor-looking sunburned pilgrims twirling prayer wheels — and squads of combat-ready Chinese soldiers. Each squad included one man carrying a 2-metre metal pole with a large hoop on the end, presumably a people-catcher.

The Jokhang Temple squatted at one end of Barkhor Square. Its construction began in 652. During the Cultural Revolution the Red Guards attacked the place and closed it down; then it was restored in the 1970s. There were queues of pilgrims prostrating themselves and slowly moving forward, between prostrations, to the entrance. Big incense fires burned outside. Meanwhile, large crowds of tourists, us included, pressed in on these intimate acts of worship. By tourists, I mean mostly Chinese tourists. Westerners like our group were few and far between. Once inside we were introduced to the things we would become familiar with over the next few days: the rancid stench of yak-butter candles and the more pleasant aroma of smoky incense, narrow passages, steep stairs, dim light, endless gold Buddhas, and the absolute crush of intermingled pilgrims and Chinese tourists. It seems that every Tibetan, and there are maybe about six million of them in the world, must visit the Jokhang Temple at least once in their life. Most of the vast number of Chinese tourists and immigrants in Lhasa visit it too.

Out of the Jokhang Temple and on to the place where we all wanted to be, the Potala Palace. We intended to go inside, but there was no need actually because it was simply fascinating to stare at endlessly from any angle. All angles were 'below'; it towers magnificently above you as it climbs to the top of its hill: whitewashed staircases and yellow buildings at the bottom, up to the massive ramparts of whitewashed apartments above, to the ochre-red top section, with dry Tibetan mountains in the background on either side. A big red flag sits on the top in the centre, and our American friend John asked why the Chinese flag was there. He didn't seem to understand that, politically, this was China. He wondered if the absence of the Tibetan flag meant that the King was not in the palace at the moment. I explained that the 'King' is the Dalai Lama, and no, he's not in the palace because he's not allowed there. He lives in exile in India. John had no idea. I think Tsering the guide was relieved that I explained this truth, as it saved him the difficulty of trying to do it in a politically acceptable way. The Tibetan reality in their homeland is awful.

The morality of the China–Tibet situation is hard to grasp. The railway means once-remote Lhasa is now just a two-day trip from downtown Beijing. The Chinese have marched into Tibet, flooding it with soldiers and settlers. They have taken over the Tibetan Buddhist religion which, since the eighth century, has been the heart and soul of Tibet. Ugly infrastructure projects have ruined a lot of the landscape. But now Tibetans have better access to health care and education than ever before. Twelve centuries of fervent Buddhist belief appear to have done little materially for the Tibetans; many of them live in appalling poverty. Their faith tells them it might be better next time round, if they behave. The Chinese don't offer the Tibetans much, but it appears to me that the Dalai Lamas didn't either.

Meanwhile, back at the Potala Palace: this was an afternoon of steps, quite exhausting at this altitude, but manageable if you took your time — broad outdoor stairways and narrow almost ladder-like indoor ones. All the rooms were quite dark, lit only by yak-butter candles. That smell! Endless apartments of different past Dalai Lamas, thousands of Buddha statues, stupas containing Dalai Lamas' bodies, and again thousands of pilgrims and Chinese tourists. Somehow, we made it close to the top.

Tsering unloaded a wealth of information on us in those dark narrow passages and broad sunlit terraces, a bit much to take in really, but you could tune out when you wanted, and go into your own reverie. After our fill of

The Potala Palace, Lhasa.

the Potala Palace, we drove back home to the Yak for a rest. Later some of us returned to the Potala for another look, unwilling to waste the opportunity. Then there was a free Welcome Dinner. Surprise! It was in last night's Dunya Restaurant next door. Yak schnitzels, which were pretty good, and chips. Chips! Depressingly, they had their clientele's tastes perfectly sorted. And we felt ready to try a beer — Lhasa beer, a weak pale brew with perhaps no hops in it at all.

Tuesday, 14 May

Lhasa

This morning after breakfast it was time to visit the massive Drepung Monastery out on the north-western edge of Lhasa. Six-hundred-year-old Drepung is one of the great university monasteries of Tibet. It's rather like a city spreading up a hillside. It was once the largest monastery in the world, and its population sometimes reached 10,000 monks in the early twentieth century; the Chinese government apparently now caps it at about 300. It has recently been a centre of Tibetan resistance to Chinese rule. In 2008 that opposition exploded: the Drepung Monastery was besieged by the 'People's Liberation' Army, with power, water and food supplies cut off, and maybe 400 deaths ensued. The monastery was closed and didn't reopen until 2013. We didn't learn any of this while we were there; poor Tsering had to stick to older and less controversial history. It seems Tibetans who speak freely to foreigners have a habit of disappearing . . .

Our entrance into the Drepung Monastery was through an ornate gateway framing the distant Himalaya. And a couple of soldiers. This was to be another day of steps, and we began the slow journey up in the company of more pilgrims and monks. Drepung was certainly not as crowded with pilgrims or tourists as yesterday's Jokhang Temple and Potala Palace. Up we trudged then, past the big sign saying 'Tibet Buddhism Academy, Lhasa Drepung Monastery Branch College'. But just who were those two tough-looking Chinese men in similar black jackets who weren't part of a tour group and didn't seem particularly religious? Looking back on it, the whole business seems creepier and creepier.

Everywhere there were rows of gold-draped prayer wheels built in, waiting for believers to brush into rotation with their hands as they passed, thus releasing thousands of prayers into the heavens. I watched an old lady with a short grey bob, gorgeous in her colourful traditional clothes, shyly turning

Drepung Monastery, Lhasa.

the prayer wheels. It was like a child going 'brrrrr' with their hand along a corrugated iron fence. I was torn between my curiosity and my invasion of her privacy. I wondered what experiences hid behind that cross-cultural gulf.

Onwards and upwards, there were lots of monks now, in maroon robes with sports shoes underneath. Some of the older ones were fiddling with mobile phones instead of prayer-wheels. The monastery-city climbed just as we did, up against the backdrop of snowy mountains disappearing into the cloud.

We eventually made it to the Debating Terrace, up near the top, a large flat paved area with a 1-metre wall around the edge and a view over the world. We got there just as an outdoor class was winding up. Perhaps a hundred robed monks were sitting on the ground around a senior figure with an extraordinary tall hat and an engaging smile. I turned to look at the truly remarkable view from the Terrace. The big monastery apartments tumbled down the hill below me, getting smaller and smaller. Somewhere in the distance, on the flat plain below us, was west Lhasa, nothing old about it at all, a mass of medium-rise buildings, presumably residential. Beyond the valley, a different world of wild mountains and snow stretching as far as the eye could see, south to, ultimately, Bhutan and India.

We broke away from all of this and went up into the Dalai Lama's apartment. Drepung used to be the residence of the Dalai Lamas, until 1649 when the fifth Dalai Lama had the Potala Palace built. We eventually found ourselves in the Great Assembly Hall, a place of total spectacle. It was festooned with luminous silk hangings and the columns supporting the roof were brightly carpeted. In that dark, smoky but vividly coloured room, rows and rows of monks, at least a hundred, sat facing each other, chanting. It was eerie and quite moving. The lowest notes were fantastic: very deep and resonant. It was dangerous to stay too long; it was all quite hypnotic.

Fortunately, the necessities of life took over. It was payday! A couple of women started walking down the rows of monks, handing out 100-yuan notes. One each and some older monks got two. It was very perplexing to see such a spiritual exercise dissolve into mere financial accounting. Brought back down to earth, we descended the hill and returned to the van.

Today was a two-monastery day; we now headed to the Sera Monastery on the northern outskirts of Lhasa, another teaching monastery on a grand scale. It also has a bitter twentieth-century history; hundreds of monks were killed during a revolt against the Chinese in 1959. These Tibetan monastery-universities are extraordinary though. They can't practise as freely as they'd like in Tibet, so they have opened large franchises in India, educating thousands to be Tibetan Buddhist monks. In exile.

Sera Monastery is now most famous for its debating sessions, in the tree-shaded Debating Courtyard. A debate takes place between just two monks, but there might be perhaps 40 or 50 pairs at it simultaneously. The debate process has rigorous rules. The defender sits on the ground; the questioner confronts him, standing. There are all sorts of ritual gestures, hand clapping, ridicule, and fake physical attacks. It is fascinating to watch and extremely noisy, especially with all those pairs going at once. The experience is visceral rather than spiritual.

Everything today had taken place in comfortable temperatures around 20 degrees, but we were exhausted by the activity and the altitude, and it was time to get back to the Yak for a rest. Eating held no attraction and I just needed to sleep.

Wednesday, 15 May

Lhasa to Shigatse

Today we were deemed sufficiently acclimatized to the altitude to begin the two-day journey to Mt Everest. Eight of us piled into our van along with Tsering and a driver. The van was okay for short trips around Lhasa, but the thought of being cooped in there for hundreds of kilometres with questionable knee room was daunting. Exiting Lhasa through the morning rush hour didn't take too long and soon we were on the Friendship Highway heading south-west through pleasant flat country, with mountains far in the distance.

This was all well-cropped farmland. The crop looked rather like rice, but surely it couldn't be at this altitude. That was a perfectly safe question for Tsering to answer; it turned out to be barley. We saw an awful lot of barley growing in the next few days. I think it's about the only thing that grows in Tibet. I learnt that barley is the staple of the Tibetan diet, not the Nepalese-style curries they serve in the Dunya. Barley flour is roasted and moistened into a dough or porridge with salty yak-butter tea, to make tsampa. That's what most Tibetans eat most of the time. As a tourist, though, tsampa was never available to me and I never saw any.

We were soon driving beside a big wide river, the Tsangpo. The Tsangpo flows from west to east through most of southern Tibet. It then performs a massive oxbow turn through the eastern Himalaya, to come out in India and Bangladesh as the Brahmaputra, mingling with the Ganges and emptying into the Bay of Bengal.

Gradually, we left the flat land behind. Popping ears signified increasing altitude, then we were on a real alpine switchback zigzag road climbing steeply into the mountains. Far below us our previous flat plain disappeared into the distance. The uphill hairpin bends were very tight, requiring lots of gear-changing from our nameless and silent driver. Rainer the German, being in the car industry, had definite ideas about how this should all be done, and he lost no time in making his dissatisfaction known. He soon went from inward seething to giving the driver a driving lesson, using Tsering as an unwilling translator. The driver meanwhile was impervious to it all. Rainer finally gave up and went back to his internal frothing. The van continued its slow grind up the hill until we reached Kamba La pass at 4800 metres, where we stopped to give it, and us, a rest. And Rainer. The view was stupendous, down the way we had come up, and on the other side down to a big deep-blue lake.

We were surrounded by arid mountains. There was no snow up here even though we were at the same altitude as the peak of Mont Blanc. Back into the rested and cooled van, and we headed on down the other side. It must be said that the roads were in fabulous condition, well-engineered and well-maintained, presumably at considerable cost. We gingerly curled down the zigzag highway to Yamdrok Tso Lake. The lake is a large body of sacred water at 4500 metres, sparkling blue surrounded by a brown lunar landscape. Supposedly, it freezes over in winter. We stopped at a stony beach where Thomas and Rainer unfurled a huge banner of a Mercedes/Chinese joint-venture car. The banner showed the car against the backdrop of Yamdrok Tso, so the boys had to get a photo of themselves with it, in the same location. A bit droll but quite amusing.

Back into the van, we cruised beside different arms of Yamdrok Tso for a while, then pulled into Nagartse, a godforsaken little concrete-block village seemingly in the middle of a desert. This was our lunch stop. At the end of the street, we climbed stairs outside a concrete-block cube, up a couple of storeys, and entered into a pleasant enough dark, cool restaurant. I sat overlooking a table of my travel-mates, looking out an open window at the concrete-block building next door.

With the exception of our Chinese-American lady and Anne and me, nobody in our group was comfortable with Asian food. I found it quite entertaining to watch them trying to find the blandest and most boring food on the menu. Chips were always a success, and spaghetti Bolognese, whatever that means in Tibet, was usually safe. Even the two Germans, who live in China, had no acceptance of or familiarity with Chinese food. I ordered a really tasty chicken and white gourd soup and felt like I'd come to the right place. It was sort of Sichuan-style food, and nothing like the Nepalese cuisine at the Dunya in Lhasa.

We continued our road trip along the flat to the head of the valley we were in then up again into the mountains, up and up, very rugged treeless country now with hanging glaciers and snow all around, till we got to the Karo La Pass at 5040 metres then on past the Nyenchen Kangsar Glacier, a

Gyantse Dzong towering over Gyantse, viewed from Pelkor Chode Monastery.

retreating glacier face right next to the road. This was a source of wonder for a while, then further on we came to the Simi La Pass, a spectacular lookout over the very long, narrow Man La hydroelectricity lake. This was a fabulous place. The water was a couple of hundred metres below us and stretched miles away into distant valleys. Up at our level, in a big gravel car park, the place was festooned with long strings of Tibetan prayer flags blowing in the quite strong wind. You could get lost in them, like clothes on a washing line. It was an absolute riot of colour against the bleak landscape and the long blue scar of the man-made lake. On and down we went, past yaks and sheep on the highway, till eventually we got to the town of Gyantse at 4000 metres.

Gyantse was a fascinating place. It used to be a major stop on the route from British India up to Lhasa, with a telegraph relay station and office. The place is watched over by Gyantse Dzong, a steeply walled fort built on a high hill at one end of the town, visible for miles and a real strategic asset when it was built in 1390. Of course, the British tried to have a finger in the Tibetan pie in the early twentieth century; believe it or not, Britain invaded Tibet from India in 1904, in an effort to counter the influence of China and Russia. Pity about the Tibetans. The British besieged Gyantse Dzong for two months, armed with artillery and Maxim machine guns against the ancient muskets and spears of the Tibetans. The walls of Gyantse Dzong were breached when a British shell hit the powder magazine, and there were possibly 5000 Tibetan casualties, compared to a few dozen for the British.

The Dzong was blown up again in 1967 by the Chinese but has since been restored to its full fourteenth-century magnificence.

We pulled up in the main street for a break. I grabbed Anne and we rushed off to the end of the street where the Dzong is. We had limited time and I wanted to make the most of it. I had seen a very romantic black-and-white distant photo of the Dzong in a book about the British expeditions to Mt Everest in the 1920s, and I was very excited to be here. We walked back to the van along the main street, looking at a busy Tibetan small town. There were women with embroidered aprons that look like carpet, and wearing traditional flat-soled shoes instead of the Western sports shoes of the big-city monks at the Drepung Monastery. Past a butcher's shop that looked like a charnel house, with a presumed yak haunch which was the biggest piece of chopped-up animal that I'd ever seen. And the vehicles! The oddest selection of three-wheelers and motorcycle-cum-trailers. And every time you turned around, there was the simply fantastic Dzong on its hill surveying the city. The lampposts were all decorated with white and gold prayer wheels right to the top, the hanging streetlamps were fake prayer wheels, and the roof lines of the two-storey-high main street façades were a colourful blast of prayer flags. Gyantse was just superb.

From here it was a two-hour drive to Shigatse along a wide flat valley through farmland. The farming was really primitive: gaily decorated horses pulling ploughs, yaks grazing, and endless barley. Finally, we made it to Shigatse, still at almost 4000 metres, Tibet's second city and quite a sprawl. We staggered into our Gang Gyan Orchard Hotel around 8.30 pm. Anne and I were quite hungry, so we went straight out to find dinner. We couldn't go with others from the group because they all wanted hamburgers.

We found a good-looking restaurant, the optimistically titled Sumptuous Restaurant, just a few doors along from the hotel, walked in, and everybody ignored us. Fortunately, we saw Tsering sitting at a table with friends of his so he got us sorted with a magnificent spicy yak soup with big fat short noodles and long thin ones too. So tasty. Just brilliant. Actually, Sumptuous. What a terrific end to a fabulous day. The altitude gets a bit upsetting towards the end of a full-on day like this, but we were managing pretty well.

Thursday, 16 May

Shigatse to Rongbuk

We were each presented with a litre-sized oxygen canister which seemed a dramatic portent of things to come. Into the van, we drove out of Shigatse and onto a valley floor of flat farmland, heading west to Lhatse, surrounded by a moonscape of rocky hills with ruins on their summits. From here we headed south towards New Tingri, up the Gyatso La Pass. This was another series of interminable switchbacks, with the driver doing his usual brake-and-clutch-stabbing routine, and Rainer quietly going crazy. Gyatso La, at 5250 metres, was our highest elevation for the whole Antipodean Express. It's on the main divide between the Ganges and the Brahmaputra. Our arrival at the flat top was heralded by a large prayer-flag-bedecked archway across the highway, with a big sign: 'You Have Entered Mt Chongolangma National Nature Reserve'. Chongolangma, of course, is the Tibetan name for Mt Everest. The Nepalese on the other side of the mountain call it Sagarmatha. But the British Royal Geographic Society, in 1865, decided to name the mountain after its past-president, Sir George Everest. Apparently, his name was originally pronounced 'Eve-Rest'.

Here at Gyatso La we should have had our first view of Everest, but it was nowhere to be seen: just grey cloud stretching away to the south. Further along on a bleak high mountainous plateau, then down continuous switchbacks for more dry farmland, sheep and, in places, lots of mud bricks out in the sun drying. There was not a tree to be seen.

Unfortunately, the highway bypassed New Tingri, also known as Shegar. New Tingri was on the itinerary of the British Mt Everest expedition of 1924 which failed only at the final hurdle, with the deaths of George Mallory and Sandy Irvine. That expedition walked and rode from Darjeeling in India into Tibet and turned west to follow the northern flank of the Himalaya to New Tingri. From there they headed south towards Mt Everest on the same route that we would take, except there was no road then.

After bypassing history at New Tingri, we soon turned off the highway onto the Zhufeng Road to Gawu La, Rongbuk and Everest Base Camp.

The road began an unbelievable series of switchbacks and zigzags, slowly ascending a long wide slope seemingly up into heaven. It made everything that had happened so far prosaic. On a map, the road looks like the doodlings of a madman, and sometimes, driving it, you could see down to all the different zigzags, connected up, coming in and out of view below and behind you. The top of all of this was Gawu La Pass, 5200 metres high, a similar height to Gyatso La, but the route was absolutely astounding. And here at Gawu La we were promised a stunning view of several of the world's highest mountains, lined up for us with Everest. 'That's where Everest should be,' said Tsering, pointing into impenetrable cloud.

We hung around for a bit, hoping for a miracle, but eventually carried on dejectedly down the other side of Gawu La, another eternity of switchbacks, to Choezom Village, where we had to clamber out of the van and into a big bus for the final 20 or 40 kilometres to Rongbuk. The Chinese government doesn't want a whole lot of private vehicles driving around Everest; the government bus was the only way to get from here to Rongbuk. Once full, the bus headed steeply up the gravel road to the Rongbuk Valley which comes straight down off Mt Everest. No views of anything; the valley walls were closed around us and the view in front was just grey cloud. The fabled Rongbuk Monastery, the highest monastery in the world at 5000 metres, appeared on the left above the road; and below on the right, the Rongbuk Guest House tent complex.

The British Everest expedition spent a night here at the Rongbuk Monastery in 1924, and received the blessing of the Abbot, before moving on to set up their base camp. The Tibetans could not understand why anyone would bother to climb a mountain.

Our bus deposited us at the tent complex. A little depression beside the road held about 20 large black-roofed tents, sitting in lines among the gravel. Inside, the tents were lined with colourful felt rugs on the walls, ceiling and floor, with wide divans around the walls for sleeping on. There were big piles of folded heavy blankets. In the middle of the tent was a stove burning yak dung.

Each tent seemed to be operated by one family who also cooked for us. Our family consisted of a husband and wife, a couple of infants, and an intermittent cast of uncles and aunts or whatever. They were all heavily dressed in wool and down, with faces of burnt leather and bloodshot sun-damaged red eyes. Some wheezed and coughed and spat, outside, and wanted to smoke inside, but a chorus of dismay from the precious whiteys, me included, soon put a stop to that. We, of course, were all feeling sick to varying extents, the altitude didn't really agree with any of us, and there's nothing better than a

Mt Everest, North Face.

Tibetan with a fag in his mouth coming into your tent to renew the misery if you'd briefly forgotten about it.

Fifteen minutes after arriving, Tsering rushed us all out to see if Everest would appear in the sunset. We slowly staggered up the brief gentle incline to the road, then on in the freezing cold towards base camp for about 500 metres, to a big sign saying we could go no further.

As if on cue the cloud filling the head of the valley stirred a bit, a few scraggly peaks of snow flickered on and off at an impossible elevation, and then there was Mt Everest glowing in all its glory in front of us. We were already in the shadow of the valley, and the massive bulk above us was stage-lit by the horizontal rays of the setting sun. Clouds scudded across the North Face, and the summit had a big cloud streaming east from the peak, against a background of dark blue sky.

The summit was 24 kilometres away, and even though we were already 5 kilometres above sea level, the awful vertical North Face towered another 4 kilometres above us. I was absolutely in awe of it all, as well as freezing cold and a bit nauseous and headachy.

The failure of the 1920s British expeditions to Everest was laid out before us in plain view. I can't begin to imagine the thoughts of George Mallory and Sandy Irvine in 1924, on the way to base camp, which they initiated. From where we were standing now, they eyeballed Everest through binoculars, trying to confirm a route to the summit: from the East Rongbuk Glacier up the North Col and onto the Northeast Ridge, which was the skyline to the left of the summit. Below the Northeast Ridge they, and we, could see the highest point achieved by the previous expedition two years earlier in 1922. Only 500 vertical metres short of the summit, George Finch and Geoffrey Bruce were pipped at the post by bad weather and good sense and turned around, bringing the 1922 expedition to a frustrating close. Indeed, we could see on the skyline of the Northeast Ridge, at a bump called the First Step, the place of the last ever sighting of Mallory and Irvine alive in 1924. No one knows how much further they got, or if they in fact reached the summit. Mallory's body was eventually found 650 metres below the peak in 1999, with a shattered right leg, a hole in the head, and eviscerated by ravens. He was otherwise perfectly frozen and preserved. Of poor Irvine, who was pretty much on his first-ever mountain climb, no trace has ever been found.

For a New Zealander, naturally, the great Everest personality is Edmund Hillary. He and Tensing Norgay, a Nepalese Sherpa, 'knocked the bastard off' in Hillary's words, in 1953. They climbed the Nepal side of the mountain which we couldn't see. Tsering hadn't heard of Hillary, but he certainly knew who Tensing was!

As the cold really started to bite, but unwilling to give up on this magical moment, we slowly walked back towards the tents, and up to the Rongbuk Monastery on the other side of the road. If we think Buddhist monks are meant to live an ascetic life, well, living at Rongbuk Monastery must be the pinnacle of asceticism. Miserable little apartments sit on a rocky slope, with not much oxygen and no warmth. There was a smallish main monastery building and a gold-topped white stupa, all in dark shadow now, with the massive mountain glowing gloriously at the head of the valley.

By now it was after eight o'clock and there was no sign of life apart from a herd of yaks who strolled into the courtyard around the same time as we did. Presumably, they were farmed by the monks, as there was no one else living around here other than the families running the tents. Goodness knows what the yaks found to eat. Once we ran out of daylight, we descended to our little tent village. It was warm inside and time for dinner: fried rice with vegetables. It made me feel a bit better. We collapsed onto the big divans to sleep. I wore everything I had. The yak dung stove had to be allowed to go out, as there wasn't enough oxygen in here for us as well as it. The weight of the blankets was crushing and sleep was a fairly fitful affair, with a pulsating headache and flushed face, and people talking outside seemingly all night.

Friday, 17 May

Rongbuk to Shigatse

We had gone as close to Everest as we were allowed, and now it was time to turn around and head back the way we came. We were at our closest point to Alaejos so far, within 8000 kilometres in a straight line, but now we would take a mighty detour to the north-east, away from Spain. We would retrace our tracks all the way back through Xining, the place of the unexpected midnight train-swap six nights ago, and Lanzhou of the famous beef noodle soup, then on to Beijing.

We started the day at 7 am, waking up in the gloom in minus 9 degrees to see the sun rise on Mt Everest. Breakfast was offered but nobody had any appetite, and we ventured out into the freezing cold dawn. The lightening sky was totally clear and the view of Everest was completely uninterrupted. Simply awesome. I staggered up to the Rongbuk Monastery again, and up a bit further to get that classic view of the soaring mountain behind the squatting low monastery buildings and their white stupa. Exhausting but rewarding.

Tsering spoke to a monk who offered to take us inside. It was quite familiar after our other monastery experiences, with bright hangings and effigies of the Buddha, but all on a much smaller scale, and no crowds of tourists — no one at all except the monk, Tsering, and about four of us. I tried to find a sense of history, thinking of the British climbers who came here 95 years ago, but I think a lot of the monastery had been rebuilt since then. I'm sure the weather takes a huge toll on the place. With mixed feelings, it was finally time to leave. Mixed feelings because being here was a highlight of my life, but I'd had enough of the headache, the breathlessness and the bone-shattering cold. So a quick visit to the unbelievably awful toilet trailer, then we all climbed into the government bus for the trip down the rough, steep gravel road to Choezom and the uncomfortable familiarity of our own van.

Soon we were on our way, ascending the still fascinating and endless zigzags up to Gawu La Pass. There was a real buzz of excitement, satisfaction and happiness in the van. We stopped again at the summit of Gawu La, and it was everything it hadn't been yesterday. Sixty kilometres to the south, laid out perfectly for us to view, was the Mahalangur Himalaya range in its astounding completeness. From

The Mahalangur Himalaya from Gawu La. From left to right: Makalu, maybe Num Ri, Lhotse right on the shoulder of Everest, Everest, and (far right) Cho Oyu.

left to right, Makalu at 8480 metres, the world's fifth highest peak. Lhotse, 8510 metres, the fourth highest. Towering Everest at 8840. Cho Oyu, the sixth highest at 8180 metres. And further to the west, in the Jugal Himalaya, the fourteenth-highest mountain in the world, 8020-metre Shishapangma. Five peaks, each above 8000 metres, glinting in the bright sunlight just for us.

There also, among the prayer flags, was a small horde of trinket sellers, very keen to swamp us and make pitiful amounts of money out of us. Tsering was always in a difficult position in these situations. He had a schedule to keep to and needed to get moving; he didn't want us to become impatient with the pestering hard sell, but he didn't want to deprive his poor countrymen of a chance for income. Tsering was a very moral man with a heart of gold.

We set off again down the north side of Gawu La. We had now achieved total gratification of all our Himalaya wishes, and last night's terrible fitful sleep without oxygen or warmth ensured that most of us slept on and off for the rest of the day's 330-kilometre journey back to Shigatse. The comparatively dense oxygen-laden air of Shigatse, at 'only' 4000 metres in the Tsangpo River Valley, was a relief to breathe. We checked in again at the same Gang Gyan Orchard Hotel on the corner, and after five days of physical closeness we were all happy to do our own thing. Anne and I went back to our nearby Sumptuous Tibetan Restaurant of only two nights ago. We just managed to keep our eyes open till the end of the meal, then that was the end of us.

Saturday, 18 May

Shigatse to Lhasa

We started the day with a rather bad breakfast of greasy stir-fried things in the sunny Gang Gyan Orchard dining room. The resident skinny cat and her mangy kitten were okay gambolling about on the floor, but not so cute when they jumped up on tables to eat scraps off abandoned plates.

This morning's task was a visit to the Tashi Lhunpo Monastery just across the road, our seventh Tibetan monastery or palace in six days. We were certainly getting our money's worth. Tashi Lhunpo is the traditional seat of the Panchen Lama, the Number Two Man in Tibetan Buddhism. Under Chinese rule the position of the Panchen Lama has been mired in skulduggery and abduction. Apparently, the real Panchen Lama was kidnapped at the age of six by agents of the Chinese government in 1995 and hasn't been seen since. Meanwhile the Chinese promote their own appointee as Leader of the Faith. He lives in Beijing. The Tashi Lhunpo Monastery was sacked by Nepalese invaders in the eighteenth century and by Chinese during the twentieth. During the last outrage, in the 1960s, the actual monastery remained intact; just most of the apartments of the 4000 monks were trashed.

So now the place, for all its magnificence, seems a bit sparse on the ground. The bleak hillside above it is a riot of long strings of prayer flags blowing in the wind, right up to the stony summit. Tashi Lhunpo was a fabulous place of courtyards, ornate doorways, paintings, drapes and a bit of squalor. Unlike the other big monasteries, it was dominated by its rather unforgiving landscape, instead of the other way round.

We got out of Shigatse and took the short cut back to Lhasa, different than the way we had come, bypassing Gyantse. In a little village Tsering fulfilled a promise he had made to Anne: Some days ago, Anne had asked Tsering if he could get her some yak wool. Now, in this village we pulled over and Tsering disappeared into a house for about five minutes, then returned with a plastic bag for Anne. Inside it was an incredibly coarse black substance rather like industrial-strength steel wool. Yak wool. This became a legend in its own right. We knew we wouldn't get it past the very tough agricultural security at the New Zealand border. So Anne got onto her sister in Germany to find someone who could wash, card and spin the wool. That person was found, and we mailed the wool from Russia to Germany. Months later it turned up in the post in New Zealand, spun into yarn. It had been softened by the addition of some other fibre, but it still only felt suitable for engineering purposes. It wasn't going to be a nice cuddly jersey for anybody.

We carried on our way, skirting the wide and quiet Tsangpo River, in clear skies through fertile well-cropped farmland. Lots of barley again. Some towns had large signs in Tibetan, Chinese and English, saying 'World Barley Town', followed by the year in which the accolade was awarded. The first time I saw one I was impressed, by the fifth I began to realize it was like the 'New Zealand Beef and Lamb Award' on the wall of every restaurant at home. We eventually drove into the familiar courtyard of our Yak Hotel in Lhasa, spilled out of the van and went our separate ways.

Anne and I were very happy to be on our own again. We had about 20 hours left in Lhasa, to do what we liked. Evening was coming on as we headed down Barkhor Street and through the famous market. There was endless amounts of tat for sale but fun to look at anyway. Out on the street we could observe the amazing evening pilgrimage circuits. There are several defined circuits around Lhasa in general and Barkhor Square in particular, which pilgrims and maybe locals too must complete. So you'd be standing in a pedestrian-only street and suddenly a tidal wave of pilgrims would materialize, in traditional dress, or smart jeans, or monks' maroon robes, and everything in between. The wave of pious humanity would engulf you in a maelstrom of whirling prayer wheels and muttered prayers.

Every so often, too, in the midst of it all, keeping pace, just to discourage antisocial religious behaviour or crimes against the benevolent state, there would be a camo-clad fully armed squad of soldiers complete with conspicuous people-catcher at the back. One of the behaviours the soldiers were there to prevent was self-immolation. It has been quite prevalent in recent years. A ghastly thought, and an embarrassing vote of no confidence in Chinese rule. I assume that the purpose of the people-catcher is to restrain the self-immolator at a safe distance while others try to extinguish him or her.

Now the muttering murmuring crowd would pass by, thankfully flamelessly, to be followed a few minutes later by another one. In my rather limited experience, I had never seen such large-scale religious activity, such scary military activity, or the potential for such horror.

The end of the day. Beijing Road, Lhasa. Potala just below the skyline.

Later, close to the Yak, we climbed up onto a high pedestrian bridge that crosses Beijing Road. Four lanes of roaring honking traffic underneath, an avenue of old four-storey buildings stretching away, an ideologically confusing conglomeration of multicoloured Tibetan prayer flags and red-and-gold Chinese flags on the roofs. Hills beyond the roofline surmounted by cumulus clouds glowing with the setting sun behind, and a few rays escaping. About a kilometre away, above the roofs, the ochre and white summit of the Potala Palace poked its head out. We stayed on the bridge for a very long time as the darkness fell about us. I wanted this moment to last for ever.

Sunday, 19 May

Lhasa to Tongtianheyan

After the van tour, this morning felt like waking up to a holiday. Outside for a final exploration of Lhasa on a beautiful warm day, around 20 degrees. A walk around old Lhasa is a confrontation with poverty and squalor. There are just so many poor people, faces wrinkled by the elements, all sniffling and snuffling from goodness knows what respiratory diseases, mumbling their prayers and whirling their prayer wheels.

Next to a dusty messy building site, we came upon a hole in the wall where a woman was selling hot chips Tibetan style: freshly fried and crisp, too hot to touch and generously smothered with salt and chilli powder. I sat down on a dirty plastic chair on the edge of the building site, watching heavily scarfed ladies sifting cement and mixing concrete, and got stuck in. The chips were magnificent and didn't last long. We headed back to the Yak through Barkhor Square for the last time — on our own pilgrimage perhaps. The big incense burners were humming, with smoke pouring out of the high chimneys. Nearby stalls were selling all sorts of different grassy-twiggy-branchy stuff to place in the fires. People were queueing up to buy it, rather the same as you would queue to buy food at the market. It's a staple of their existence.

Back at the Yak, we rendezvoused with Tsering as planned, and he drove us to Lhasa Station for the train to Beijing. We passed through some of new Lhasa, concrete, grey, sterile and rather bereft of people. We didn't yet have actual tickets for the train; we were still in the care of GTT and Tsering had our vouchers. He queued up with our passports and Tibet Permit to get our tickets. Of course, he needed his special tourism worker ID to be allowed into the queue, which was only for passengers and therefore, by definition, not Tibetans. Tibetans aren't allowed to go anywhere. They aren't allowed to have passports because they don't need them. Then it was goodbye for ever to dear, lovely, gentle Tsering. I told him where we'd been and where we were going, but

Barkhor Circuit, Lhasa.

he wasn't really interested. His current life, and the next ones, were in Tibet.

Down long escalators to the platform where our immensely long green train waited for us, with its two big colour-coordinated diesel-electric locomotives at the front. Then into our carriage, and the familiar minute scrutiny of Tibet Permit, passports, visas and tickets. It seems you can't even go out of Tibet without the Permit that attests that you're allowed to be there in the first place. Comfortably seated in Train Z22, Lhasa to Beijing. Forty and a half hours, 3750 kilometres: our longest train trip so far, and the second-longest of the Antipodean Express.

Once again, we were in a four-berth cabin. Our sole roommate at the start was an uncommunicative pig of a man who spent most of his time shouting into his mobile phone. The train left fifteen minutes late, at 4.30 pm, as we had to wait for a Xining to Lhasa train to clear the single track first. We slowly cruised out of Lhasa and the smooth comfort was indescribable after the cramped, bumpy, jerky van. The mountain scenery outside the window soon became fantastic and then the whole place got a dusting of light snow. Magical.

Around 7 pm we checked out the dining car, which we had missed on the train from Xi'an. What a revelation! We entered through a green forest of vegetables growing in pots, into a modern, clean, roomy restaurant car with gorgeous sunset light flooding in through big windows. Red plastic flowers next to the windows made it quite civilized and almost romantic. There was an à la carte menu in Chinese, and also, much cheaper, the pre-cooked stuff in a plastic tray which the trolleys hawked in the corridors, and we had eaten a week ago, on the way to Tibet. The waitress suggested the plastic tray, easier for her, but fortunately, this time Anne insisted on challenging the real menu with Google Translate. We ended up with spicy chicken, stewed brisket and rice, cooked to order and delicious, and a new kind of insipid barley beer. It was a very enjoyable meal in unbelievable surroundings. 'Dinner in the diner, Nothing could be finer, Than perhaps a cup of tea in southwestern China!'

Out the window about a kilometre away we could see the raised earthworks of a new high-speed railway being constructed parallel to this one. This railway was an extraordinary engineering feat when it was built thirteen years ago, and already it was about to be superseded by something much faster and more advanced.

Back to our cabin then, and at depressing Nagqu we lost our loud, unpleasant roommate; he was replaced by two men who at least said *ni hăo* pleasantly, but that was about it. After Nagqu we began the laborious grind up to the Tanggu La Pass at 5200 metres again, on the boundary of Tibet and Qinghai Province. There was snow on the ground and miserable-looking yaks huddling together as snowy mountains seemed to grow higher. The animals disappeared as the place became more and more desolate in the dusk.

We flattened out for the run alongside bleak Tsonag Lake in the last of the light, then we pulled the curtains and went to bed. One of our new roommates turned out to be an appalling snorer; but general train-noise and earplugs got me to sleep soon enough, though Anne was less lucky. She was in the top bunk, and he was in the opposite top bunk, and she felt that he was snoring right into her ear. I was asleep before Amdo and was not aware of Tanggu La Pass. And I wasn't there for Tongtianheyan either, which must have been around midnight.

Monday, 20 May

Tongtianheyan to Liangzhen

Not much to say about a long day in the train that took us through the provinces of Qinghai, Gansu, Ningxia, Inner Mongolia and Shaanxi. We woke up somewhere after Golmud, after a pretty bad night's sleep thanks to the chronic snorer and general altitude unease. We had been up over 5000 metres again during the night and now we were still around 3000. Fruit for breakfast, bought at Lhasa Station, and two-minute noodles for lunch. As the day wore on time slowed down. The scenery had lost its Tibetan Wow factor and we were feeling the lack of privacy, so there was lots of sleeping and reading.

Anne also took the opportunity to get some violin practice done. Three months is a very long time for a professional musician to stay away from her instrument; the muscle-memory and the finger-callouses deteriorate quite quickly. It wasn't practical to bring a fragile, temperature-sensitive violin in a slightly bulky case on a trip like this. Anne's solution was to have a mini practice-violin built by a luthier friend: just a violin neck and finger-board with a cut-down scroll and four short strings stretched between the original pegs and a little low bridge at the other end. No body or resonating soundbox, so hardly any sound. You couldn't really 'play' it, but it was tunable and perfect for practising fingerings, fast or slow, with the left hand, and, for the right hand, there was a cut-off frog (bottom end of the bow) mounted on a pencil.

From time to time you'd hear a quiet tinkle-tinkle and that would be Anne keeping her violin chops in shape. A brilliant invention, the whole deal was only 23 centimetres long, weighed 100 grams and took virtually no space. After practice, we pulled into Xining again, a chance to get out and stretch the legs but fortunately no train change this time. Then Lanzhou about four o'clock where we entered new territory, heading north-east through Inner Mongolia towards Beijing instead of south-east to Xi'an.

Sometime before dinner we visited the dining car to photograph the menu so we could take it back to our cabin and apply Google Translate to it at our leisure. Dinner was the highlight at the end of an otherwise dull day. What

Goodbye Tibet.

we wanted to order wasn't available anyway, so the menu exercise was a bit futile, but we did okay. Back to our cell, where sleep didn't come easily after a day of lying around, and the snorer was on shockingly good form; I needed a sleeping pill and earplugs to keep him at bay. Liangzhen happened in the middle of the night, but I was certainly not awake for it.

Tuesday, 21 May

Beijing

We woke at first light to discover that our two roommates had left us sometime in the early hours and had been replaced by one uncommunicative, but fortunately non-snoring, girl. If you board a Chinese sleeping car in the middle of the night, you just climb into the previous person's sheets.

There is a large clear area at the end of each carriage, next to the toilets, with basins, taps and mirrors, where everyone communally performs their morning rituals: the first spit of the day (as noisy as possible), teeth cleaning, make-up, hairstyling if you are under 25 and male, maybe even a couple of loud phone calls. This was the end of the journey. We rolled into Beijing West Station absolutely on time at 8.33 am and staggered out onto the strangely unmoving platform. Straight into a taxi to the Penta Hotel in Chongwenmen, some distance from Beijing West. I had chosen this hotel because we would be leaving early tomorrow from the old Beijing Station in the middle of town, not Beijing West. The Penta was within walking distance of Beijing Station.

Now we had the rest of the day to spend in Beijing. This was our third visit together, and we'd already seen many of the things Beijing has to offer: the Forbidden City, the Great Wall, the Summer Palace and much else besides. Today was just a 24-hour staging-post to catch our breath, and the train to Russia. Our special treat would be a visit to the pandas in the Beijing Zoo.

But, first, some business. Arrival in Beijing signified the end of our relationship with GTT. From tomorrow when we step onto the Trans-Mongolian Express, until arriving in Paris in fifteen days' time, all our travel arrangements would be looked after by our man Matt at Monkey Business here in Beijing. Three weeks ago, sitting in Lemongrass Restaurant in Ho Chi Minh City, I had received from Matt a photo of a flagpole out the front of Beijing Station. We should be at that flagpole at 6.30 am tomorrow to meet a guide who would issue us with tickets for the 7.27 am Trans-Mongolian Express to Irkutsk. So the first thing we should do right now was to rehearse the walk from the Penta to the Beijing Station flagpole, to find the way, make sure it was actually walkable, and see how long it took: 20 minutes. Just fine.

We caught a bus to the zoo. The trip was about an hour, and it turned out to be an interesting look at suburban Beijing. It was hot, 32 degrees according to the read-out on the bus, but pleasant enough.

Alighting from the bus, the zoo was on the other side of a busy multi-lane arterial route and access was through a shopping mall under the road. There was a pharmacy in the mall and Anne thought this would be a chance to buy something to get her bowels moving. The appalling state of the toilets throughout Tibet had put her alimentary canal into lockdown. Even with the help of Google Translate, the task of communicating to a non-English-speaking pharmacist's assistant that you wanted a laxative, and why, was a bit hilarious. Finally, Anne got the message across, we thought. The woman insisted that she understood, and produced a little packet of pills, without a single Latin letter on it. I suggested that Anne should take the first pill now, but she fortunately insisted on waiting till we got back to the Penta that evening, when we would have time to apply Google Translate to the instructions on the packet. That turned out to be a very good idea. After more Google-induced hilarity, we worked out that Anne had bought an antidiarrheal, the opposite of a laxative. Which could have had dire consequences. I'm sure there's a lesson there somewhere . . .

Meanwhile, at Beijing Zoo, we entered Panda Heaven. Well, for us anyway, perhaps not for the pandas. Quite large animals, doing nothing much other than eating bamboo, it seems they spend fourteen hours every day doing just that. We stayed for a long time, making the most of an approximately once-in-a-lifetime opportunity. Finally, we took the metro, much faster than the bus, but less interesting, home to Chongwenmen and the Penta.

I think that any trip to Beijing is incomplete without Peking duck for dinner, and that was the next thing on the agenda. We had a booking at a very upmarket Peking duck restaurant, the Jing Yaa Tang, in the equally upmarket Sanlitun area of Beijing. Telephone reservations for non-Chinese speakers in China being doomed to failure, we had got a Chinese violin-colleague in the NZSO back in Wellington to arrange it for us. She had contacted a friend in Beijing, by the name of Winter, to make the booking. So now we descended to the metro again, in the thickest rush-hour. We dived in and after half an hour in a human sardine-can we were spat out into the rarefied air of Sanlitun.

Into the restaurant we went, all dark wood panelling, soft lighting, high ceiling, tables of the Beijing young-and-beautiful and lots of expats. Half-cooked ducks hung up by their broken necks were on display in a well-lit glass

Peking. Duck.

cabinet. Of course, the German maître d' had no record of our booking, but he found a table for us anyway. Although they couldn't find our reservation, they did find our pre-order for half a duck. Weird. Anyway, we got our Peking duck, and a couple of other dishes. It was all very delicate, refined and subtly excellent. Perfectly crispy roasted duck skin, with the flesh too, which isn't always the case. All sliced up next to the table by a razor-sharp cleaver-wielding virtuoso in a little black hat. Then you select a few julienned vegetables, smear on a bit of sweet bean paste, and wrap it all in a little thin pancake. That's Peking duck. Don't expect to get a full stomach on it! This was fabulous eating, fabulous drinking, and very interesting people-watching. All that was left to do was to pay an awful lot of money. We later learned that Winter, who made our booking, received a phone call from the restaurant during the evening, wondering where we were . . .

We left the restaurant and walked around Sanlitun in balmy air, in the company of well-off young things doing much the same. Groovy boutique fashion shops, night clubs, a fascinating English-language bookshop with impressive coffee-table-sized books. I couldn't help casting my mind back to the grim headachy squalor of Shigatse and wondering if this was the same country. Well, according to the Tibetans, no, of course it's not. And Sanlitun isn't just different than Tibet, it's also more international, less Chinese, than Hangzhou or Xi'an.

When we had had enough of being on our feet we took the metro back home. Chongwenmen had seemed perfectly comfortable earlier in the day, now it seemed decidedly down-market after the delights of Sanlitun.

Wednesday, 22 May

Beijing to Erenhot

We were up and out the door before 6 am, ready for our rehearsed walk to the station as Beijing woke up. We got to the forecourt flagpole rendezvous with time to spare, and found our anxious but efficient young Chinese minder, courtesy of Monkey Business. Including us, he was putting six tourists on the Trans-Mongolian Express today. There was quite a crowd gathering in the big expanse of the forecourt, with the Chinese-Socialist-Realist twin clock towers of Beijing Station rearing above us. The excitement of people about to depart on a very long train journey was palpable all around us.

The Trans-Mongolian Express goes all the way to Moscow, but we would be going only to Irkutsk in Siberia on this train, 2680 kilometres, in 55 hours. Tickets finally in hand, our guide ushered us through security into the big spacious station.

In its day Beijing station was the largest in China; it must have been very impressive when it was built by a desperately poor nation in 1959. Up to a waiting area in one of the upstairs galleries; then we had plenty of time to kill and an opportunity for breakfast. There were a few Chinese fast-food joints and I found a really good breakfast of wonton soup with seaweed, and a youtiao fried bread stick, the same as a Vietnamese quẩy. This was followed, I'm sorry to say, by a KFC cappuccino which lived down to expectations. The station filled up with people as the day got going, but it was never as busy as the frantic monster-stations in Nanning and Guangzhou. The big blue departure board had only five departures in the next three hours on it. Soon we were on our way down the escalator to our next long green train, Train K3 to Moscow. Each carriage had a plaque on it in Russian Cyrillic, Chinese characters, and Mongolian Cyrillic, announcing Beijing-Ulaanbaatar-Moscow.

Once, eleven years previously, I had been at the same station and I saw one lonely carriage with this plaque on it. Back then it seemed impossibly exotic, and I wondered if I would ever get into a carriage like that.

Now here we were, it was happening, and it was hugely exciting. The whole business of boarding was very easy and stress-free compared with most of our experience so far. What a luxury to have a two-berth cabin to ourselves again. Two bunks, one above the other, lying transversely across one side of the cabin, and one comfortable seat next to the window, facing the bottom bunk, on the other side. Dark wood-veneer panelling, plush red velvet curtains and trimmings, and carpet on the floor. In an odd arrangement, we shared a bathroom with our neighbours. A basin and handheld shower, no toilet, situated between our cabin and the one next door, with access from each cabin. Our neighbours were friendly Cheryl and Yip from Singapore, fellow Monkey Business travellers.

The thumps and exclamations of people moving into their accommodation for the next few days carried on for a while, so I took the opportunity for a stroll to the front of the train for a look at the locomotive, a big brutal blue rectangular block of a thing with pantographs up to the overhead wires. The railway through north-east China and Mongolia to Ulan-Ude on Russia's Trans-Siberian line was opened in 1956 just after I was born. The entire route through to Moscow basically follows the ancient Beijing to Moscow tea-caravan route which used to take 40 very bumpy horse-drawn days, until 1898 when the western half of the Trans-Siberian railway was opened.

Everybody on board, doors closed, we left Beijing Station exactly on time at 7.27. One and a half hours ago the streets had been almost empty and now we were in the midst of the rush hour, crossing on viaducts over six-lane highways packed with traffic: cars, taxis, vans and a few buses.

On my first visit to Beijing, or Peking as we were still calling it then, in 1975, there was, if anything, more congestion, and it was all bicycles. Absolutely no private cars, a few belching heavy trucks, and carts pulled by strange two-wheeled tractors with long handlebars stretching back to the driver sitting on the cart. And a slow-moving dark jungle of black bicycles ridden by an army of people in Mao jackets. You could express your individuality through your choice of dark blue or dark grey.

This time around, the city of 20 million thinned out surprisingly quickly and after 30 minutes we were following a large river, maybe the Sanggan, westwards through a mountainous gorge with countless other railways and bridges and tunnels going off in every direction. Being a lot older than the Tibet railway, the engineering scars on the landscape had had time to soften, no longer the gaping wounds on the land that constantly confronted us on the Tibet train. This was all stunningly beautiful, and in fact the Great Wall was supposed to be an occasional companion up on the rugged skyline. I looked and looked but I didn't see it. Standing, or sitting on the fold-down seats, in the corridor was a treat because the top half of the windows on that side opened all the way, and it was great for photography and a breeze in your hair.

Passengers had been allocated seating times in the dining car, and our lunch sitting was rather early at eleven o'clock. Another pleasant modern Chinese dining car, this time with quite a few white faces, offered rice, celery, and a sausage and carrot stew. It wasn't very good. There was no choice, so there were no problems with translating the menu. The lunch did, however, have the added attraction of being free. Well, included in the ticket-price anyway. After lunch we went back to the luxurious privacy of our own room, and we arrived in Datong for our first stop at 1.30. We didn't get out. Datong was originally set up as an army outpost, part of the Great Wall defences, and much more recently its claim to fame was as one of the last manufacturing centres in the world for steam locomotives, until 1989.

The landscape gradually flattened out to a dull green featureless plain, the precursor, I guess, of the Gobi Desert. Distant mountains disappeared. Camels materialized in the nearer distance. Dusk came on around seven o'clock, the sky and the desert slowly transformed to a glowing orange; it was quite magical.

The sun disappeared in a blaze of glory at ten to eight and we pulled into Erenhot, the Chinese stop for the Mongolian border, in the very last of the dusk at 8.15. Chinese emigration was a relaxed and rather slow affair. We stayed on the train and pleasantly polite Chinese border guard ladies visited each cabin and took our passports away. The train stayed put for an hour or so after that.

Mongolia was just a couple of kilometres away, but there is a major impediment for trains travelling from China to Mongolia. In China the rail gauge is 1.435 metres, known as Standard Gauge throughout China, Europe, America and Australia. Standard Gauge was invented by the Englishman George Stephenson, of 'Rocket' fame, in the early nineteenth century. He measured the axle-widths of 100 horse-carts and came up with an average of 4 feet 8.5 inches. That's 1.435 metres. In Mongolia they use the slightly wider Russian gauge of 1.52 metres, meaning a train cannot run unchecked between China and Mongolia. The solution is to separate all the carriages, lift them up individually, and replace all the wheels with new sets of wheels, or bogies, of the correct gauge for the new country.

Now the train was broken up into separate units, an endless process of

The Trans-Mongolian Express awaits new bogies at Erenhot on the China–Mongolia border.

bangs and jerks, starting in the distance and getting closer and closer like a thunderstorm, until it was our turn. Then we were shunted into an enormous brick shed, and banged back and forth until we were lined up with the big hydraulic jacks that would lift us up. All this took forever, and other carriages in front and behind and parallel beside us went through the process at the same time.

A train crew member appeared in our corridor, tapping and patting the carpet, looking for something. Dissatisfied, he lifted the carpet, still to no avail. He went away and brought back someone else, and a heated discussion ensued, with lots of pointing into our cabin, at me. Our man then indicated that he wanted to lift the carpet of our cabin, and I started to wonder what was hidden under there and how uncomfortable Erenhot jail might be. But all was in order, and under the carpet, just inside our cabin door, was the vertical pivot of the wheel bogie, the pivot which allows the four-wheeled bogie to turn for curves in the track. The pivot had to be unlocked with a large wrench, then lifted up out of the bogie, through our cabin floor.

The pivot was a thick, greasy steel shaft maybe a metre long. Our man couldn't budge it at first, and it took lots of banging and grunting before he finally got the shaft out and laid it neatly and greasily on the carpet, making a nice mess in the process. Someone must have gone through the same business at the other end of the carriage, and now we were ready to be jacked up, leaving the disconnected bogies on the ground. After a whole lot more banging and jerking, we were imperceptibly slowly lifted about one and a half metres.

I could watch all this in better detail by observing the carriage next to us. The abandoned bogie sat on quadruple tracks which accommodated both gauges. The Chinese bogie was wheeled out and then a Mongolian one was wheeled in and lined up. The carriage was then gently lowered. Everything had to be perfectly aligned for the pivot shaft in our room to be reinserted; that took several attempts and lots more banging. Then once we were down the brakes and other linkages all had to be reconnected. After an age we were shunted back out into the night, and the train was jerkily and noisily reformed. The Chinese dining car had been extracted, and a new Mongolian one inserted. A white Mongolian diesel-electric locomotive with blue and red stripes was connected to the front. Finally, it looked like we might be ready to go soon, and passports were brought back to our cabins.

Thursday, 23 May

Erenhot to Sukhbaatar

Departure from Erenhot happened at 1 am. This was our farewell to China; we'd been there for seventeen days and it was our longest stay in one country for the whole Antipodean Express. We had covered about 11,000 kilometres in Chinese Railway trains. As the train pulled out of Erenhot, I was in bed already and fell asleep immediately. Half an hour later I was jolted awake by the train pulling into Zamiin-Üüd for Mongolian immigration. Fortunately, the Mongolian officials came onto the train because I was barely awake. I managed to hand over both our passports to a strapping policewoman in a blue uniform, short skirt and knee-high black leather boots. She was quite an apparition to wake up to, but nothing compared to the next one, a heavily made-up female soldier glamour-puss in immaculate combat fatigues and desert boots. It was all like something out of a James Bond film.

My first impression of Mongolia in daylight was the bleak coal-mining town of Sainshand at 20 to 7, viewed from the comfort of my bed. Rough block one-storey houses were scattered among the brown dirt, surmounted by a vast bright blue sky. After leaving Sainshand we were really in the Gobi Desert, with periodic herds of camels for diversion. We rolled into the new Mongolian dining car looking for breakfast. What a spectacle! The place was festooned with stunningly intricate wood carvings of birds, flowers, clouds, a balalaika-type-thing and lots of abstract pieces. We were almost first into the diner, but it soon filled up with mostly Chinese or Mongolian groups. The only breakfast option was decidedly un-Mongolian: scrambled eggs on toast. Coffee was a plastic Edeka jar of instant granules, '*Gut und Günstig*', a teaspoon, and a cup of hot water out of the samovar. No milk, just a jar of white *Kaffeeweisser* powder. All of this while watching the Gobi Desert trundle by. Things could have been a lot worse.

Back in our cabin, we carried on with Mongolian sightseeing which was certainly engrossing. The Gobi stretched for ever in every direction, but unlike the red centre of Australia it had a tinge of dull green. A gravel road seemed to follow this part of the railway so that, even though there was no traffic, it didn't feel totally remote. There was a ramshackle telephone line all the way. And for a desert, there was a surprising number of animals: cattle, horses and camels. Quite a few dead ones, too. There must have been some humans because every so often there would be a solitary yurt next to an empty animal enclosure, with a four-wheel-drive parked alongside. After lunch the countryside became noticeably greener, with more cattle and horses and fewer camels. And now solid houses as well as yurts. But just a house was obviously not enough; every single house had a yurt next to it.

Eventually, we hit the outskirts of Ulaanbaatar, with its multicoloured roofs under depressing dun-coloured hills. We passed by new suburbs of yurts between wooden fences. Yurt suburbs were apparently the result of recent mass migration from the Mongolian countryside into Ulaanbaatar, causing all sorts of social, infrastructural and real-estate imbalances. We pulled into Ulaanbaatar Station which was a two- and three-storeyed affair of light grey granite blocks with teal blue-green trim. This was a long stop of about an hour, so we went into a small supermarket to get some fruit, and Mongolian tugriks out of an ATM. Access in and out of the rather small station complex was easy, unlike Chinese stations, so we went out the other side and stood in Ulaanbaatar for a while, even if it wasn't quite the centre of town.

Mongolia is actually a centre of Tibetan Buddhism, despite the geographical separation, and Ulaanbaatar once had a population of 30,000 Buddhist monks. Mongolia gained independence from China in 1911 and in 1921 it became basically a Soviet satellite with all the repression, terror and destruction that implies. In fact, Ulaanbaatar translates as 'Red Hero'. Since the early 1990s the country has been a democracy with a rapidly changing social, economic and political make-up.

We headed back to the station to see how things were going there. Our locomotive had been removed and replaced with two huge two-tone blue diesel-electrics with white stripes and white galloping Mongolian horses painted on their flanks. They looked like something out of the early 1960s or perhaps a Tintin book. Soon it was time to get going, and we left heading north, uphill with good views back down to the messy, scrappy collection of yurts and houses with an ugly high-rise centre, filling the window.

Mongolia has a population of only 2.8 million, and nearly half of them live in Ulaanbaatar, which also has the distinction of being the coldest capital in the world. We had really left the Gobi Desert behind by now and were rolling through green, undulating farmland, with trees, houses, creeks, flocks of sheep

Mongolian dining car.

and herds of cattle. Dark hills appeared in the distance as the sun got lower.

We decided not to be late for dinner and turned up at the dining car at five o'clock. There was plenty of variety on the menu and it looked good. Anne ordered a chicken dish, and I ordered another chicken dish, completely different than hers. When they arrived, they were identical and bore no relationship to what we thought we had ordered, apart from the actual chicken. When I politely questioned her, the waitress insisted, through a confusing absence of mutual language, that the two dishes were not at all identical and were in fact exactly what we had ordered. I left it at that. Where else were we going to eat?

Meanwhile, things were rocking and rolling along. I had assumed that the extra wide Russian gauge would give us a more comfortable ride, but despite the comfort of the Chinese train, the Russian bogies gave us a ride quality that was positively agricultural. Nonetheless, as night fell I was asleep soon enough. We pulled into Sukhbaatar on the Mongolian side of the Russian border just before 10 pm and I was woken by a knock on the door. It was the next procession of beautiful Mongolian babes in military marching-girl uniform, ready for another instalment of the now familiar customs and emigration ritual. It was all quite pleasant, but they did have a really good look under the bed and in the bathroom. They took the passports away as usual and returned them with a smile an hour later; I had to stay vaguely awake for that, and we departed from Sukhbaatar, and Mongolia, at 11.10.

Friday, 24 May

Naushki to Irkutsk

Russia! At a quarter past midnight, we were at Naushki, on the Selenga River which flows into Lake Baikal. Naushki, 5 kilometres from Mongolia, was the Russian border post. Russian border police came onto the train, and I opened one eye, hoping that, as in Mongolia, I could just hand over the passports. But, no, they insisted I wake up properly and get my feet out of bed and onto the floor. The Chinese train crew came through and removed all the roof lining in the cabin and the bathroom. Then the Russian police came back and had a thorough look in the roof cavities and under the bed and the single seat. Next came the Russian customs woman in overalls to look through our luggage and ask lots of questions. Finally, a gorgeous little low-slung Beagle sniffer dog paid us a visit, just to make sure. We were on our way again in our reconstructed cabin at 2 am.

Four hours' sleep, then consciousness reappeared around dawn. The train was still in the Selenga Valley, surrounded by wet farmland, approaching Ulan-Ude. At Zaudinsky on the outskirts we joined the 9200-kilometre Vladivostok to Moscow Trans-Siberian Railway. The train pulled into Ulan-Ude, our first taste of Russia in daylight, and we had an hour or so here, so we went out to have a look.

Right now, with the sun not really up yet, it was cold out on the platform. The read-out said 8 degrees, but it felt much worse than that. We were standing at what seemed like a rather small station building, quite beautiful in pastel blue, marooned in an enormous ugly marshalling yard. There were a few other empty passenger trains drawn up but not going anywhere, and a lot of goods wagons sitting around. I climbed up to a long footbridge over the rails for a good view. Now there was a big red and grey RZD Russian Railways electric locomotive on the front of our train. From up on the footbridge, I could see the high-rises of the town nearby and low hills in every direction in the middle distance.

A couple of young women came by, presumably walking to work, warmly and neatly dressed with big coats, hats and long boots, and immaculate make-up. They were both quite Asian to look at, but their clothing wasn't. The nearby buildings and houses looked European. After six weeks in Asia, it unexpectedly felt like I had suddenly arrived in Europe. Actual Europe was a long way away beyond the Urals, but this place, superficially at least, felt European even though the only two random citizens I had seen were Asian. The apartment buildings could have been in Germany.

Actually, this place has its own non-Russian social structure. Ulan-Ude is the capital of the Buryat Republic, a Federal District of Russia. The Buryats are Mongol people who inhabit the area roughly between Mongolia and Lake Baikal. There are half a million of them; traditionally they are Tibetan Buddhists, but shamanism is also a main feature of their spiritual life.

After an hour at Ulan-Ude Station we moved on, following the Selenga River through wetlands and, suddenly, Russian birch forests. We passed through Selenginsk, founded in the seventeenth century as an Orthodox missionary outpost to convert the heathen Buryats. Now it's known for its wood-pulp factories which drain their waste into the Selenga, slowly poisoning the once-pristine Lake Baikal.

Meanwhile, breakfast in Russia seemed a good idea. Not only did we have a new Russian locomotive, we also had a new Russian dining car. It was rather a shock to the eyes, with high-backed narrow seats upholstered with screaming bright red and white vinyl. Still the visuals didn't upset the bowl of porridge; not oats but wheat or some other grain, with a block of butter melting in it. The culinary adventure continues!

We caught glimpses of blue water through the trees, and then, around Boyarsky, the whole of Lake Baikal was laid out before us in all its glory as we travelled west along its southern shore, a vast expanse of cold blue water and blue sky. The western shore ahead of us, crowned by the snow-covered Primorsky Range, dissolved into distant haze. Likewise, the eastern shore behind us, backed by the epic-looking Khamar-Daban Range. The inland ocean disappeared north over the horizon; the two shores meet each other at Nizhneangarsk, an unbelievable 600 kilometres away to the north. Baikal is a place of superlatives. The deepest lake on earth, it holds a fifth of the world's freshwater supply. That water is crystal clear and safe to drink, except, not by coincidence, in the southern corner where we, and our railway, were.

We reached Mysovaya, which was the end of the line until 1904. Before then, it was here that the train would drive onto the icebreaker *Baikal* for the voyage across the lake to Port Baikal on the west shore. The train would then drive off onto the Port Baikal wharf and continue down the Angara River to Irkutsk and ultimately Moscow. The 4200-ton icebreaker, incidentally, was built in Newcastle-on-Tyne and brought in pieces to Port Baikal by train. Soon after Mysovaya, we passed the Baikalsk Cellulose and Paper Combine, which until recently pumped its untreated chlorine-contaminated waste water straight into the lake. With this plant, and the pulp factories at Selenginsk, the future of Lake Baikal doesn't look very bright.

Now we ran right beside the lake, in a sort of seaside reverie for the next couple of hours. It was endlessly fascinating, with the expanse of water to our right and snow-covered rugged mountains behind birch trees to our left. We were on the turn of the seasons. The birches had remarkable soft-looking bright green fresh growth contrasting with their white skin, and there were still occasional spots of leftover ice on the shore. Baikal, with the same surface area as Belgium, freezes over completely in winter, to a depth of about a metre, and the thaw was only four to six weeks ago.

Having made it onto the actual Trans-Siberian line, traffic had picked up considerably. There was a regular procession of goods trains going in the opposite direction to interrupt the spell. Meanwhile we were getting our first look at Siberian rural life: lots of 'charmingly' decrepit small wooden houses, with ornate eaves and shutters painted brightly in pastel blues and greens. At some stage there was even our very last Tibetan Buddhist stupa, gleaming gold and white, festooned with strings of prayer flags, gazing over the implacable lake. The Buryats and their Buddhism are a large part of the human presence around Lake Baikal.

Sometime around midday the train stopped at Slyudyanka for a bit. Slyudyanka is a nothing of a place right next to the water, but its station is a marvel, built entirely out of locally quarried pink marble. In 1904 the so-called Circumbaikal Line was completed, filling in the extremely challenging last 260 kilometres in the middle of the Trans-Siberian Railway, around the southern shore of Lake Baikal. This made the underpowered British icebreaker the *Baikal* and her smaller sister ship the *Angara* redundant, and vastly improved the Trans-Siberian as a strategic resource. The completion of the Circumbaikal Line was such a momentous event that it was commemorated in the building of a magnificent marble station at Slyudyanka.

This makes an interesting comparison with The Last Spike in the North Island of New Zealand. Just four years after the completion of the Circumbaikal Line, in 1908 the Main Trunk Line between Auckland and Wellington was completed at a very remote place in the central North Island called

Manganuiōteao. This was also a very significant event, and the Prime Minister, Sir Joseph Ward, hammered in the celebratory polished silver Last Spike before a large crowd of gentlemen in suits, coats and hats in disgusting weather. So a silver spike then, but certainly no grand commemorative marble station.

Meanwhile, back in Russia, the Circumbaikal Line still exists, but much of it is no longer part of the Trans-Siberian Railway. We left Slyudyanka and a while later came to Kultuk at the south-west corner of Baikal, the junction for the old Circumbaikal Line and the Trans-Siberian. Here the Circumbaikal continues to the right, eastwards to Port Baikal. We, on the Trans-Siberian, turned left, up over the Primorsky mountains on a steep, winding newer railway built in 1955, north-east to Irkutsk. So we departed Kultuk, much like an airliner, going

Downtown Irkutsk.

straight up with lots of banking turns affording views of scrappy little Kultuk beneath, and the awe-inspiring lake and mountain ranges beyond. Soon Baikal disappeared behind birches and larches and hills, but this wasn't farewell; we would be coming back here tomorrow, to spend a couple of nights at the lake. The railway over the Primorsky Range was quite steep with lots of tight bends and tunnels, and this really slowed us down, till eventually we cruised down the other side to the valley of the wide Angara River.

Big bridges and high-rises announced the arrival of Irkutsk, the Paris of the east, population 620,000. We pulled in dead on time at eight minutes to three o'clock. This was farewell to the plushly carpeted and curtained Trans-Mongolian Express. We were met on the platform by Sasha, a driver from Monkey Business. Sasha took us in his late-model Japanese SUV over the big river and into the old centre of town, to the charmingly quaint Marussia Hotel, a four-storey wooden cube. Our upstairs corner room was big and airy, with polished wooden floor and whitewashed rough-cut and split weatherboard walls. It was very rustic but totally comfortable with all mod-cons and a view out the window, of what seemed to be the trendy young busker-ish part of town.

We were out on the street in no time; we wanted to get to a bank before they all closed, to get rid of 40,000 Mongolian tugriks. We found a bank and waited a while, for a teller who told us they don't do foreign exchange. Fair enough, we tried another bank with a sign saying '*Kassa*' and the promise of *Wechsel*. We took a number and sat for ages, watching the bureaucracy creak

along. When it was our turn, I put the tugriks on the counter. The woman there just laughed and shooed us away. Oh well. It had been interesting to watch life in a Russian bank.

So, out into the curious semi-1960s time warp of Irkutsk, amid avenues of three- and four-storey brick and stone buildings with beautiful façades in pastel colours. Footpaths held a few people looking not much different than Pravda photos circa 1965, and the rutted potholed main streets were packed with, by now, rush-hour traffic. Lots of Soviet-style trolleybuses and a few distinctly communist-looking old light trucks and vans rattled by but hardly any Russian-manufactured private cars. I saw two or three clapped-out ancient Ladas, and no new ones, just thousands of Japanese cars and a few European ones. And lots of standstill congestion.

Beyond the centre of town, the city seemed to consist mostly of dilapidated and rotting one- and two-storey wooden buildings. They were exceptionally beautiful. The paint had peeled off most of them, except for some brightly coloured window frames and lintels. Windows and roofs were invariably crooked, and there was some gorgeous filigree decoration on eaves. I was shocked to discover that most of these ancient-looking quaint wooden ruins were only about 100 to 120 years old, the same age as our wooden house in Wellington. I initially thought these places were just neglected, but it must also have to do with Irkutsk's fierce climate, which ranges from minus 30 degrees to plus 30.

We walked past a huge Lenin statue with the usual admonishing finger, looking just like the one in Hanoi but without the football players, and ended up on a wide, empty, windy embankment next to the flat expanse of the Angara River. Despite the blue sky it was getting uncomfortably cold, and we headed back to the Marussia. In the square below the hotel, people were a bit younger and didn't look so much like *Izvestiya* propaganda. But this was Russia: An immaculately dressed and barbered dude sat busking Bach on his cello, through a sound system, while his decidedly dodgy-looking leather-clad and patched comrades sat on and stood around an equally immaculate Harley-Davidson motorcycle.

By now a Russian dinner seemed a good idea. Our man Sasha had suggested a nearby restaurant run by a brother or a cousin, assuring us it was real Russian food, so we took him at his word and found the place, and it didn't disappoint. Back in time again, perhaps to the 70s. Unknown Russian movies were playing on an old tube TV in a wooden cabinet with the sound muted. The clientele seemed mainly Russian, both white and Asian, with perhaps a table or two of tourists. We were entertained by a fabulous slim young Putin-lookalike male singer in a pale blue shirt who gave us endless renditions of Russian songs to a pre-recorded accompaniment on a reel-to-reel tape deck. From time to time tough-looking Russian men would lead their plump blonde wives through a couple of rounds of the dance floor. The combination of the music and the setting were magical, and I had to pinch myself.

Then came the food! A trio of smoked salmon, lightly salted herring, and raw omul, a small white fish famously endemic to Lake Baikal. Washed down perfectly by a Nerpa vodka. Nerpa is a freshwater seal, also a Lake Baikal endemic, which fortunately only gave its name to the vodka. My main course was a substantial mushroom soup in a bowl of hollowed-out bread dough which had been baked to something resembling salty terracotta. And between us a bottle of gutsy Georgian wine. We ate reasonably well in Russia, but this, our first meal, was absolutely the best, and one of my most memorable meals ever. I don't know what the restaurant was called, but *spasibo* Sasha for the recommendation. This day was an endless eye-popping adventure and now all that was left was to get back to the Marussia and sleep.

Saturday, 25 May

Irkutsk to Listvyanka

The next two days would be spent at Lake Baikal. The place is too special to just whisk past it on the way to Spain. Activities and accommodation were difficult to explore and confirm from New Zealand, even online, and part of the attraction of Monkey Business was that they offered various things to do and places to stay, which we could pick and choose from. I really wanted to take a ride on the Circumbaikal Line from Slyudyanka to Port Baikal. That can only happen on a few days a week but, yes, Saturday was one of them. And, unbelievably, the Saturday train was hauled by a steam locomotive! A ride along the shore of Lake Baikal in a steam train. That was almost beyond comprehension from my viewpoint in New Zealand.

We had a very early start with breakfast down in the bowels of the Marussia, so early that we were the only ones there apart from the young Komsomol boy serving us. This was a real Russian breakfast: weird pancakes made from a fermented sweet batter, a porridge of strange seeds, boiled eggs, sausage and cheese. We certainly weren't hungry by the end of it.

A driver from Monkey Business picked us up. His name was Sasha, just like yesterday's driver, and he delivered us back across the Angara to the station where our train was waiting for us. It barely met the definition of 'train', as there was only one carriage, with an electric locomotive to haul us the 130 kilometres back over the Primorsky Range and down to Slyudyanka where the steam locomotive would take over. The carriage was full, with some 40 Taiwanese, a number of Mandarin-speaking Russian guides, two Germans, and Anne and me. So about the same as a tourist train in New Zealand then. We set off on the Trans-Siberian line again, into the Primorsky forests, and two unexpected pre-packed substantial breakfasts were plonked in front of us. I didn't want to appear ungrateful, so I did my best.

When we got to Slyudyanka I went up onto an overbridge to have a look around. The electric loco was uncoupled and driven away, and our solitary carriage stood forlornly at the platform across from the commemorative pink marble station building. The station had lots of separate lines; our carriage occupied one, the next one over was left vacant, presumably for through-trains, and all the other lines were occupied by stationary goods trains.

It seemed a quiet Saturday morning at Slyudyanka, but that was soon shattered by the almost forgotten but instantly recognisable moan of a steam whistle, and there was our steam locomotive approaching slowly from the distance. As it got closer it became apparent that it was in fact two steam locomotives. Absolutely massive ones. Two L-class 2-10-0s coupled back to back, or tender to tender, 103 tons each. These behemoths were going to haul our solitary carriage to Port Baikal. Each of Russia's 32 Railway Administrations keeps ten of these monsters in running condition as a kind of strategic reserve. Nobody was in much of a hurry, and it took forever to get the locos coupled to our carriage. Meanwhile, a long passenger train pulled in on the vacant through-line, possibly the Moscow-Vladivostok Trans-Siberian, judging by the number of obvious foreigners who jumped out to ogle and photograph these unexpected simmering and steaming giants from the past. I felt quite smug to be an actual participant rather than a passing spectator.

We got going and crawled off into the Baikal shore wilderness towards Port Baikal, 95 kilometres away to the north-east. We passed through Kultuk, the same as yesterday, but this time we left the Trans-Siberian and turned right on the Circumbaikal, heading east on the north shore around the coastal cliffs. From here to Port Baikal was a miracle of engineering, created in just two and a half years from 1902 to 1904. The project was originally shelved in 1893, considered too difficult. However, completion of the rest of the Trans-Siberian, and the resultant increased patronage, meant that the two British icebreakers could not cope with all the people and freight, and the Circumbaikal Line had to be reconsidered and built.

Most of the route was originally cliff-face plunging into the lake, punctuated by valleys and rivers. All the work was done manually by a huge workforce of convicts, free Russians and immigrants. There were usually about 13,000 men working on the project. Just about every metre required the land to be modified with tunnels, galleries, bridges, and revetments to prevent landslides. Eventually, there were 41 tunnels and hundreds of bridges. All of this was finished with stone facings of incredible skill and beauty, much of it the work of Italian stonemasons. People worked 16-hour days tied to rock faces through the Siberian winter. Now we could cruise along it in steam-hauled comfort, our two steam monsters chuffing lazily, leaving an ecologically unsound but nostalgically beautiful wake of coal smoke behind us.

L-4253 next to a tunnel on the Circumbaikal Line with Lake Baikal and the Khamar-Daban Range behind.

The train stopped regularly at little stations or interesting bridges or tunnel façades just to have a look. Some of the stops were quite long and the locomotive crews always took the opportunity to dismantle various bits of steaming apparatus on the sides of the locos, stand around them concernedly, make adjustments, then reconstruct them. Meanwhile there were always amazing stone works, wooden bridges and views to consider.

Right on midday, two substantial aluminium-foil packages of hot lunch were dropped on the table in front of us. Chicken breast stuffed with capsicums. As if two breakfasts weren't enough!

At Kirkirei-3 Cape the railway couldn't be built around the outside of the cape, so we puffed through it in a tunnel then stopped at the other end. Right next to us, closer to the lake, was the portal for an earlier tunnel abandoned in 1914 and well worth exploring. Next to the portal was a cast-bronze plaque covered in blistered paint with raised text in Russian and English:

> **Tunnel No. 18 'Kirkirei-3'**
> Rail joint of the Great Transiberian Railroad was commisioned here on 13 september 1904.
>
> Head of the National Railways Department M. I. Khilkov drove in the last spike.

So the completion of the Trans-Siberian Railway was not only commemorated with a marble station at Slyudyanka, but Prince Khilkov also hammered in a last spike too. Right here at Kirkirei-3 Cape. No mention of silver, perhaps the pink marble of Slyudyanka Station was enough of an extravagance. I don't know this, but I'm sure the Kirkirei-3 spike was installed before a large crowd of gentlemen in suits, coats and hats in disgusting weather. I feel a little special to have stood at both Manganuiōteao and Kirkirei-3 . . .

Around three o'clock we stopped at a wooden platform near the mouth of the Bolshaya Shumikha River. Here was the Eastern Siberian Railway's Tourist Resort House Shumikha: five or six bare weatherboard houses with corrugated iron roofs in a lush green valley. This turned out to be the dinner stop, and all the Taiwanese sat at long wooden benches on the terrace of the main house. Nobody told us anything, and we were still full anyway. We were content to look around for a while.

The view back across the grass to the locomotives on the embankment, quietly simmering, with the infinite blue lake behind, was like a dream. A blue sky with a few clouds, and it was quite warm. An old man in a shed was selling tourist junk but he also had tempting home-made pastries and I had to try one even though I had no room for it. Then after a while one of the guides came and said we were supposed to be over there with the others having dinner. Something about we'd been catered for, so we had to have it. So we sat in the pine pavilion for a magnificent borscht and salad. The following main course chicken was wrapped in foil, like the lunch chicken, for us to take, because the train was about to leave for the next 2-kilometre leg of the journey. Just as well, because neither of us could have fitted the chicken anywhere.

We trundled off into the late afternoon, and stopped somewhere to walk through a curved tunnel, again built to replace the one next to it whose access had disappeared in a rockfall way back in the day. Inside the bare stone tunnel, I found the truth of the local climate: stalactites of ice growing out of the walls. They were completely dry; no melting in here.

Sometime around 6.30 we pulled into Port Baikal. This was the end of the line. The Circumbaikal Railway stops here, although it didn't used to. Port Baikal is on a point on the lake where the Angara River flows out of Baikal and north-west to Irkutsk and ultimately into the mighty Yenisey River. Between 1904 and 1955, the railway carried on past Port Baikal, around the bend into the Angara River valley, and followed the Angara for the 60 kilometres to Irkutsk, thus completing the continuous railway from Vladivostok to Moscow. However, in 1956 the Angara was dammed, raising its level enough to permanently flood the railway between Irkutsk and Port Baikal. A small realignment was not possible; the Trans-Siberian would have to go somewhere completely different between Irkutsk and Lake Baikal. So that is when the new railway over the Primorsky Range, between Irkutsk and Slyudyanka, which we took this morning and yesterday, was built. And that is why the Circumbaikal Line has become a historical cul-de-sac curiosity.

One more thing about the Circumbaikal Railway: as I pointed out in yesterday's story, before the Circumbaikal was built it was necessary to take the icebreaker *Angara* across the lake and carry on by train from there. In 1904 at the beginning of Russia's disastrous war with Japan, many thousands of troops were sent east from European Russia on the Trans-Siberian Railway. Brilliant. This was exactly the sort of strategic purpose it had been built for. But when the troop trains arrived at Port Baikal in minus 40 degrees, the frozen lake was beyond the capabilities of the two icebreakers, which could not move. And the Circumbaikal Railway was not ready yet. But the troops had a war to get to, so rails were laid directly onto the 1.5-metre-thick ice and a trial locomotive was sent out on them, with entirely predictable results.

Consequently, thousands of soldiers had to walk — not march! — across the 40 kilometres of ice to Tankhoy on the southern shore, a journey that took 17 hours. The locomotives were dismantled and dragged over in pieces by huge teams of men and horses. Not to mention the heavy artillery and ammunition. At Tankhoy the trains were reassembled and continued their journey to war. This was obviously unworkable and Prince Khilkov, the government minister in charge of building the railway, put immense pressure on his workforce and somehow the Circumbaikal Line was ready to transport the army by the end of 1904, about a year ahead of schedule, and about a year before Russia inevitably lost the war anyway.

Now, here we were at Port Baikal in the cooling evening sun. The place looked quite interesting in a depressingly faded industrial debacle sort of way. I would have loved to have spent some time exploring the historical treasure trove of Port Baikal, but all 50 of us had to rush through the industrial wasteland to queue up for the narrow gangplank onto a ferry which would take us across the source of the Angara River to the bigger town of Listvyanka on the other side. The ferry was quite old, very narrow and maybe 30 metres long, with a bravely fluttering Russian tricolore. At Listvyanka we struggled with our packs up the gangplank, and I wasn't quite sure what would happen next. The wharf seemed to be some distance out of town and I didn't know where to go. Up a short steep road following the Taiwanese and their guides, and suddenly there was our next Sasha with a big sign saying 'Gregory Bill'.

Three cheers for Monkey Business! And, yes, he really was called Sasha. He was a quiet unassuming balding man of my age, with reasonable English. He loaded us into his clapped-out old car and drove us to our accommodation for the next two nights, the Usabda Demidov Guest House. He dropped us off and said he'd come back to pick us up in the morning. The Usabda Demidov was a two-storey motel-type place close to the lake. Our upstairs unit was all pine, perfectly warm and comfortable. At the front of the property, on the road next to the lake, was the Listvyanka Club, which was reception, restaurant and bar.

Out on the waterfront in the sharp crisp light and very long shadows of the Siberian evening, we could see that Listvyanka stretches narrowly for kilometres along the one lakeside road. We followed this road for a long way, with the Khamar-Daban mountains to the right, glinting distantly but hugely in the setting sun, beyond the blue desert of the lake. We ended up in a quite built-up area, with a market, across the road from a sort of port, with lots of fishing and pleasure boats pulled up bow-on to the beach. The boats were all similar: long, narrow, with large boxy superstructures — quite attractive, and, I think, quite unique to Baikal.

Our lovingly prepared foil-wrapped chicken dinners had been at blood-temperature for about five hours, so they regrettably ended up in a rubbish bin. Now the market here seemed the best place to eat. Most of the waterfront, and especially the market, was occupied by stalls selling omul, the endemic fish. The fish were all sold whole, gutted, and held open by toothpick-like skewers. They were available salted or smoked. Once under the cover of the market, the smell of smoking fish was impossible to resist. As is often the case in this sort of situation, Anne and I ate completely differently. I went to an omul stall and the friendly young woman wrapped up a beautiful smoked omul, about 30 centimetres long, in paper for me. Anne went to another stall run by a couple of criminal-looking young dudes who sold her a disgusting dry plov, as in pilaf, which might have been cooked two days ago, and a pork shashlik which could have been cut from a plank of wood.

We took our food upstairs, with a couple of beers, and sat outside on the roof of the market in the sunset, dwarfed by the immensity of Lake Baikal. Anne's food was as I have described; my omul was a sensation. Beautifully moist, eaten off the bone with greasy fingers. I got into a fabulous mess. The very slowly setting sun finally dipped below the hills around 20 to 9, and by now it was very cold. Well wrapped up, basically in our Mt Everest uniform, we strode out, back along the waterfront to the eventual warmth of our Usabda Demidov, draped on the outside with magical-looking fairy lights.

Sunday, 26 May

Listvyanka

The morning sun, the crisp air and the lake gave everything an unbearably dazzling brightness. Yesterday evening's Sasha turned up at 10.30 to take us and three others for a walk in the forest above the lake. The others were Yip and Cheryl, the two Singaporeans whom we had first met at Beijing Station four days ago, and Sasha's wife Margarita. Yip and Cheryl were a pleasant surprise; we hadn't expected to see them again. Margarita was along for a Sunday walk, and to help her husband communicate with us because her English was excellent. She would also provide the all-important lunch that she had prepared.

Off we strode, straight up the side of a steep hill into pine forest. This was real Eurasian forest, nothing like New Zealand rainforest. There was a clean understorey with a literal forest of straight, vertical, well-spaced trunks, all on a crazy angle because we were on the steep hillside above the lake. Despite the bright spring sun, it was only 10 or 12 degrees, but soon the warmies had to come off. It was exhilarating to be outside and exerting myself. I really relished getting stuck in with the legs on the slope and getting my lungs going. From time to time there were clearings with stunning views back across the lake, and occasional Baikal boats leaving wide wakes on the barely rippled surface.

The forest was fantastic. Some parts were dominated by birch trees with brand-new growth and the most beautiful clean, white, fresh paper bark. Elsewhere it was all larch and Siberian pines, which supply the very patient with pinenuts. Pinenuts are a bit of an industry around Baikal, with a number of stalls down at the lakefront selling the nuts, and, very expensively, the oil.

Sasha had a story of misery concerning pinenuts. We in the West regard Gorbachev's perestroika years, around 1990, with great positivity, but for many Russians perestroika meant poverty, unemployment and social dislocation. Sasha and his brother both found themselves unemployed, an unheard-of affliction in the Soviet Union, and there was no welfare system for this new class. So they headed off into the wilderness of the Khamar-Daban mountains on the other side of the lake, to harvest pinenuts for a few seasons. It was backbreaking work with only a small cold tent to retreat to at the end of the day. The only 'labour-saving' device was a massive sledgehammer with a long handle, to bash the tree trunks and hopefully dislodge a few nut-bearing pine cones.

At some point we had climbed high enough, with a stupendous view before us, then we carried on with a long steep descent. Presumably, this descent could have continued for kilometres beneath the water. Lake Baikal occupies a rift valley that's 25 million years old. The rift was originally 9 kilometres deep, and still is, but over the years it has filled up with about 7 kilometres of sediment so now the water is only 1.7 kilometres deep. It really is an extraordinary place on so many levels. Meanwhile, we eventually crashed to beach level and fortunately stopped. It was lunchtime. The beach was a narrow pebbly affair. Notwithstanding our long climb up and down, we were just one bay east of the long stretch of Listvyanka, though it felt like we had the whole of Siberia to ourselves. Even with the blazing sun, it was still cold and there was a nasty bitter wind coming from the north, so we had to scout around to find a little headland to shelter behind.

From out of Margarita's pack came an absolute cornucopia of prandial joy. Tender chicken thighs, carrot plov, red capsicum and home-baked dark Russian bread. And a thermos full of tea. We washed all this down with a wee dram of Sasha's exceptional moonshine. Next came the most delectable sweet poppyseed pastry. This was nothing like a commercial guided excursion at all. It was more like Margarita and Sasha had invited us into their living room, which just happened to be the most fantastic lake on earth, for lunch.

We had to get going back to Listvyanka, by a lower route this time. The forest here took on a magical aspect, with purple, yellow and blue spring flowers everywhere. The white birch bark kept the light in the forest quite bright and the flowers looked just superb. Directly below us was the water, and through the dappled birch leaves you could see right through the water to the lake bed. Presumably this was not the bit that is 1.7 kilometres deep.

We made it back to the east end of Listvyanka and it really was Sunday afternoon party-time on the lake front. Here there was a pleasant reserve area with lots of grass, and public barbecue pavilions, all full of families and friends having fun and pretending summer had arrived. It was freezing! This was the first we had seen of Russians having fun. To my eyes most of the young men, crew-cut, rough, tough and macho, looked like convicted drug dealers. The

Listvyanka. Russians enjoying the Rite of Spring at the beach in Siberia.

seaside footpath was crammed with promenading people of all ages and even the rocky beach had people lying on towels among the pulled-up boats. So this was a day at the beach in Siberia.

Here Anne and I said goodbye to Sasha and Margarita and the Singaporeans. We wouldn't see them again. We loitered around the east end for a while, deriving endless fascination from watching Russians and their lake.

At some stage we called in on the Nerpinarium. The Nerpa is Baikal's endemic freshwater seal, and that makes them unique. They have enormous deep black eyes and puzzled faces, so everybody loves them. There isn't really a convenient way to see them in the wild. If we wanted to see them, we had to swallow our pride and our consciences and watch them perform for us in captivity. Which we did, at great expense. They were gorgeous and cute and well-trained, and their trainer girl treated them nicely, but I felt uneasy about the whole business. Extraordinary animals, though, and nobody knows how they got into Lake Baikal in the first place.

After that we came out into the freezing cold late afternoon and took the long walk home on the waterfront to Usabda Demidov, dinner at the Listvyanka Club, and bed.

Monday, 27 May

Listvyanka to Nizhneudinsk

Plans for today involved a drive to Irkutsk and a guided walking tour of the city for a couple of hours, then in the afternoon we would be on the Trans-Siberian Railway for three days, to Vladimir. Into the very first Sasha's SUV, then at the source of the Angara we turned right to follow it to Irkutsk. The journey took about an hour on a straight but rather rutted highway, with very little traffic. The Angara River was on our left all the way. The original, now flooded section of the Trans-Siberian Railway between Port Baikal and Irkutsk was beneath the water on the other side of the river. Finally, into the traffic madness of Irkutsk and Sasha took us straight to the big ugly Soviet-era Angara Hotel to leave our luggage and meet our guide for the afternoon. Fortunately, she was a woman, so she was called something other than Sasha.

She was Kristina and was a great guide and a lovely person, with an exhaustive knowledge of Irkutsk and Russia. She had very pragmatic and honest views on Russia and Russian politics, and world politics, and didn't mind sharing her opinions on the current Russian leadership. So the afternoon wasn't just a look at sights and explanations of them, but real three-way discussions as well.

We had already seen some of Irkutsk a few days ago, and I had read a bit about it. The town was founded in the 1650s as a tax-collection outpost and developed as a stop on the tea caravan route from China and a fur trading centre. Irkutsk really hit its stride in 1826 when the Decembrists arrived. In December 1825 a group of intellectuals, military officers and noblemen attempted a coup d'état in St Petersburg, during the brief interregnum between Tsars Alexander I and Nicholas I. The coup failed and most of the rebels, now labelled with the month of their uprising, were exiled to Irkutsk to serve out their sentences, first in labour camps but ultimately in much more lenient conditions. They brought their wives, families, upper-class values and intellectual refinements with them. Soon many built palatial homes for themselves. Thirty years later Tsar Alexander II pardoned all the surviving Decembrists and invited them back to the west, but many chose to stay in the social and intellectual bubble they had inflated for themselves: the Paris

Girls waiting to stand guard by the Eternal Flame at the Irkutsk War Memorial.

of the east. Hence the architectural beauty of much of old Irkutsk, despite a catastrophic fire in 1879. The two world wars didn't come here, though the city was shelled and damaged in the Civil War after the Revolution.

We walked exhausting miles with Kristina, endlessly talking politics and social issues with the occasional stop to look at a building here or there. We got to the War Memorial, in front of the grandly palatial Regional Administration Building. This place was astounding. In the centre of a large flat area was a low brown marble plinth smothered in flowers, with the Eternal Flame in the midst of it. Standing rigidly to attention, unmoving, two on each side of the Flame, were two schoolgirls and two schoolboys in military uniform. There was obviously a roster, so no one had to stand to attention for too long; there was a nearby bench under a tree with four more soldier-schoolgirls joking and gossiping, waiting their turn. I didn't see any spare boys though. I just couldn't imagine Kiwi school kids doing this. I asked cynical Kristina if she did this as a kid, and did she respect the sentiment, and she said that everybody does it, and everybody respects it, and she did too. Wow.

Sasha drove us to the station. There was a fair bit of hanging around, but it was interesting to watch Russians waiting for trains, and eventually our big red and grey train pulled in with no fuss or fanfare. I had obviously lost my sense of direction; the train appeared to me to be going the wrong way. There is no train called the 'Trans-Siberian Express', as the name refers more to the railway. Our train was Train 001, the Rossiya, travelling from Vladivostok to Moscow. We would be taking the Rossiya to Vladimir, 4025 kilometres in 72 hours, the longest section of the whole Antipodean Express.

The Rossiya stops for quite a while at big cities, so there was no rush to get on the train. We were travelling SV class: Spalny Vagon, just two berths per cabin, one on each side, with a table between. It was rather spartan and bland after the rich woodgrain and drapes of the Trans-Mongolian, but perfectly pleasant, clean and comfortable. Evidence of previous occupants had been expunged and fresh clean blankets, sheets and pillowcases were neatly folded on the bed. No bathroom or toilet, they were down the end of the carriage. Actually, there was a basin under the short liftable table, but the table quickly becomes a repository for camera, toothbrush, books and everything else, so using the basin is more trouble than it's worth. At 3.55 pm on the dot, we were off.

What was on offer now was endless Siberian farmland and forest, which alternated with sleep. Around 6.30 we thought we'd have a look at the dining car. The sign on the door said 'Closed until 7.00 for technical reasons', but we opened it to have a look anyway, and the lady told us to come in and sit down. The place was virtually empty: just two young bandits across the aisle drinking beer and discussing something earnestly. Later a Chinese couple walked in, but that was it, while we were there. Meanwhile the Siberian sunset flowed past outside. The west side of each birch tree positively glowed orange. We sat watching for ages, then went off to bed as the Siberian darkness came on.

We stopped at Zima, which means 'Winter'. Later we passed through Kuytun, which means 'Cold' in the Buryat language. Sometime around midnight there was a brief stop at Nizhneudinsk, whose reputation revolves around sawmills, swamps and sandflies. It's also uncomfortably close to Tunguska where there was an enormous explosion on 30 June 1908. The blast was heard 350 kilometres away, and the flash was seen in the dark throughout Europe. Two thousand square kilometres of forest were flattened. Fortunately, no one was living there. It was probably caused by a piece of a comet exploding in the atmosphere. I pulled the blanket up over my head for protection and went back to sleep.

Tuesday, 28 May

Nizhneudinsk to Tatarsk

After Nizhneudinsk the Rossiya carried on through the night, through Gulag territory. In the very early hours we stopped at Tayshet, which was a major transit camp for Gulag prisoners heading further east, and a permanent camp for the poor souls who began the construction of the Baikal Amur Mainline. The BAM is a huge project, begun in the 1930s, and opened in 1989. The railway starts here at Tayshet and goes north then east through uninhabited wilderness and permafrost, past the north end of Lake Baikal and on to Sovetskaya Gavan, on an inlet of the Sea of Okhotsk facing Sakhalin Island. That's just north of Japan. Sort of parallel to the eastern Trans-Siberian Railway but a few hundred kilometres further north. It's either a 4000-kilometre white elephant or a major strategic transport asset, depending on your viewpoint.

Of more immediate importance to us was our first time-zone change since entering China at Pingxiang three and a half weeks ago. Clocks back an hour, sometime in the middle of the night. Right now we were in the midst of a big change of direction. Irkutsk is on the same longitude as Singapore, and the same time zone, though five and a half thousand kilometres due north. Now, after Tayshet, we were gradually veering from north to west to cross Russia towards Europe, and the time-zone changes would come thick and fast (the map of Central Russia is reasonably accurate, but its projection means directions are somewhat distorted).

Time-zone changes are a very new phenomenon for the Trans-Siberian Railway. From its inception until only about a year before our travel, the Trans-Siberian operated exclusively on Moscow time across seven time zones from Moscow to Vladivostok. Irkutsk is five hours ahead of Moscow, so it used to be that you'd walk into Irkutsk Station at 3.30 in the afternoon for your train to Moscow, and the clock on the platform said 10.30 in the morning. Quite confusing when you're a long way from Moscow, but the Russians stuck with that system for over a century, only changing the railway to local time in about 2018.

Somewhere east of the Yenisey River the sun started to come up and we passed through the charmingly named Uyarspasopreobrazhenskoye, which even the Russians prefer to abbreviate to Uyar. At about eight in the morning we rolled onto the new bridge across the mighty Yenisey River which flows for 5000 kilometres from Mongolia to the Arctic Ocean. I was starting to understand that everything in Russia is on a huge scale. The longest river in New Zealand is 400 kilometres . . .

On the other side of the Yenisey, we pulled into Krasnoyarsk, a city of a million people, for a long stop. A leg-stretch, but not enough time to go out of the station. Russian railway platforms are alleged to have old ladies selling food on them, but I never saw one on the whole trip, which was a big disappointment. Nothing exotic for breakfast then. Krasnoyarsk became an important military-industrial centre during World War II, safely out of reach of the Germans, manufacturing equipment to replace materiel destroyed in the fighting. It retained its industrial focus after the war to the extent that foreigners were not allowed into Krasnoyarsk until 1989. Time to get going then, and we left Krasnoyarsk, past endless rows of dachas and freshly cultivated allotments. Plenty of beetroot and potatoes growing here, while we shot past in our cabin eating a breakfast of fruit from the Irkutsk supermarket.

We continued with a restful train day of reading, sleeping and fabulously monotonous Siberian scenery: forests, farms, silver birches, villages, cities, silver birches. It was all magnificent.

At once appropriately named Taiga, no longer in the midst of dense snowy taiga forest, we passed the junction of the 80-kilometre branch line to Tomsk, where in 1993 Russia had a post-Chernobyl reminder when a plant for reprocessing radioactive waste blew up, releasing radioactivity over a few villages. Sometime after Taiga the clocks went back another hour, making for quite a long day. Still more birch forest, and very wet ground. Sure enough, we passed through Bolotnaya. The name translates as 'Swampy'.

Time for another dinner in the diner. According to the menu, my Meat Soup Solyanka Merchant Style, served with sour cream and greens, consisted of beef, sausages, smoked brisket, cucumbers, onions, tomato paste, olives and lemons, and contained 342.0 calories, Proteins 4.18, fats 6.88, carbs 4.0. All for 450 roubles. The following Beef Stroganoff Russian Style with fried potato was similarly dissected and accounted for. The proprietress brought the complete selection of available wine to the table and lined them up like soldiers. The only red was a strange sweet Georgian, so we went for that.

Suddenly we pulled into big Novosibirsk Station. We climbed down onto

Train staff and shadow on the platform at Novosibirsk.

the platform in brilliant evening sunlight and warmth and mingled in the large crowd. There were a couple of kiosks on the platform selling bananas and lots of plastic packets of junk food, but certainly no babushkas with steaming hot saucepans of pelmeni and potatoes. I wouldn't have had room anyway, I suppose. Novosibirsk, at one and a half million, is Russia's third-largest city and, as the name suggests, it's relatively new. It didn't exist before the Trans-Siberian Railway reached it from Moscow.

Soon it was time to go; we climbed back on board and straight out of Novosibirsk and onto the 125-year-old Great Ob River Bridge, almost a kilometre long. As the sunset and the Ob disappeared behind us, we entered the interminable wet, swampy, mosquito-ridden Baraba Steppe, punctuated by a few small towns through the night. By midnight we must have been somewhere near Tatarsk.

Rail
Road
Kara Sea
Barents Sea
NORWAY
FINLAND
WESTERN RUSSIA
Gulf of Finland
St Petersburg
ESTONIA
LATVIA
LITH.
Kirov
Perm
Tyumen
Ishi
Suzdal
Moscow
Vladimir
Nizhny
Novgorod
Yekaterinburg
Orsha
Smolensk
Minsk
Chelyabinsk
Ufa
BELARUS
Brest
Volga River
UKRAINE
KAZAKHSTAN
Rostov on Don
ROM.

Wednesday, 29 May

Omsk to Kirov

Omsk arrived at about 2.15 in the middle of the night. This was another million-plus city and lots of people were getting on and off the train, and making as much noise as they possibly could, right outside our cabin door. I gave up on sleeping and lay there angrily, eyes wide open, staring at the ceiling, waiting for it all to stop. As we left Omsk the noise died down and the eastern sky started to glow about 3 am. Sleep continued to elude me; tonight's ride over the rails was particularly rough.

A hundred and fifty kilometres after Omsk, at Nazyvayevskaya, the clocks went back yet another hour. Nazyvayevskaya has the distinction of being our closest point to the Kazakhstan border, just 50 kilometres away to the south-west. Soon after Nazyvayevskaya came a stop at Ishim, where Alexander von Humboldt was almost shot in 1829, while carrying out a geological survey for the Tsar. Humboldt, standing on a hill surveying Ishim through a threatening theodolite, so upset the local police chief that a detachment of armed soldiers was sent up the hill to keep an eye on the supposed German spy.

We got out of Ishim in one piece and carried on to a stop at Tyumen at 8.30 am. Tyumen looked like a big city with bad rush-hour congestion. It had been raining, but at Tyumen Station we emerged onto the platform into freshly cleaned air and bright sunlight, under the sternly watchful eye of our provodnitsa.

The provodnitsa is the carriage attendant. There are two for each carriage. Usually built like a tank, the provodnitsa projects a fearsome visage and inspires instant obedience. Among other things, she is responsible for the frighteningly complex coal-fired samovar at the end of each carriage. It never goes out for the whole journey. The odd thing is that as you exit your modern steel carriage on this electric train, you walk through a cloud of coal smoke. The provodnitsa is also responsible for keeping the stainless-steel toilets at surgical-grade cleanliness. The moment you vacate the toilet, she's in there

Somewhere on the Sylva River in the Urals.

with gloves, brush and squirt-bottle. I wouldn't mind having a provodnitsa at home. And she did smile occasionally. The element of surprise made her smile thrilling and disarming.

So, a few minutes of fresh air among the crowds on the Tyumen platform, then back on board through the steely gaze and the coal-smoke cloud. Tyumen was a major transit point in the Gulag system. The Tyumen Forwarding Prison was legendary for its horrific conditions of misery. The excrement buckets certainly never saw a provodnitsa. At least, when the railway reached Tyumen in 1888, the prisoners could finally come here in cattle wagons. Before that they were herded over the Urals on foot. Time to leave Tyumen then. Just out of town we passed the frontier pillar between Siberia and the Urals. The old road used to come this way too; it seems that almost every prisoner who ever went to Siberia came past this sentimental and tragic landmark. The commemorations of misery continued 40 kilometres further on at Yushala, where the sailors who mutinied on the battleship *Potemkin* in 1905 were shot.

Now we started to gradually wind our way into the foothills of the Urals, and there was heavy industrial development on the left which led us into Yekaterinburg, a city of 1.4 million, at 1.40 in the afternoon. Yekaterinburg is where Tsar Nicholas II and his family were murdered by Bolsheviks in 1918. The city was subsequently renamed Sverdlovsk after Yakob Sverdlov, the party official in charge of the murders.

Sverdlovsk gained further notoriety in 1960 when a CIA U-2 reconnaissance aeroplane flying near the city at 68,000 feet was shot down by a Soviet missile. Proof that the USSR did not hold the monopoly on callous disregard for human life, was provided by the US Congress. Congress criticized the pilot, Gary Powers, for: first, not blowing up the plane and, second, not committing suicide, and allowing himself to be captured 45 kilometres south of Sverdlovsk, instead. Sverdlovsk reverted to the name Yekaterinburg again in 1992.

After a half-hour break, we moved on from Yekaterinburg, up into the gently underwhelming Urals where we soon passed the obelisk at Pervouralsk that marks the nominal border between Asia and Europe. We had transited Asia from south-east to north-west and now here we were in Europe. I must say it didn't feel any different. On through dense green and white forests and along very attractive big rivers, with isolated towns and gold onion-domed churches. We rumbled on through Kungur Station, a big little place on the beautiful broad Sylva River, and down the west side of the Urals to Perm. From 1940 to 1957 Perm was named Molotov after the foreign minister. The cocktail appeared about then too, named by the Finns during their Winter War with the USSR. As the train left Perm we crossed the kilometre-long bridge over the mighty inland waterway the Kama River, which flows into the Volga. Names that sounded comfortingly familiar, though I had never been here before.

As we went to bed in our now well-worn and messy cabin, we put the clocks back another hour. The next place of interest was Kirov, around midnight. Another city of exiles, it takes its name from the high-ranking communist Sergei Kirov, who was assassinated in 1934. There is a junction at Kirov, and a branch line goes north to Kotlas, where Alexander Solzhenitsyn's *One Day in the Life of Ivan Denisovich* is set. After all this depressing history at midnight, one benign fact about Kirov stands out: it was the northernmost point of the entire Antipodean Express route, at 58°36' north.

Thursday, 30 May

Kirov to Suzdal

During the night the Rossiya blasted through Sherstki where the clocks went back yet another hour and we had finally made it to Moscow time. Our Rossiya approached Nizhny Novgorod from the north, across a long high bridge over the Volga, Europe's longest river. There were broad but fleeting views of the kilometre-wide river and the expansive city in the morning sun, and that was it.

During the communist years Nizhny Novgorod was renamed Gorky, after its most famous son, Maxim Gorky, who was born there as Aleksey Maksimovich Peshkov in 1868, into a childhood of misery and bitterness. He later reflected this by taking on the pseudonym Gorky, which means Bitter. So it's quite appropriate that in the Soviet era Gorky was a closed city, infamous as a place of internal exile. Andrei Sakharov was exiled there from 1980 until Mikhail Gorbachev set him free in 1986. Gorky was reopened, and reverted to its original name, in 1991.

We pulled into Nizhny Novgorod Station exactly on time at 7.34 am. Hoping to find something for breakfast on the platform, we jumped off the train for a quick look. I took my wallet, but we left the train tickets and passports in our locked cabin. We weren't going far, and we knew we had fourteen minutes. A quick search of the platform came up with no food at all, so we set off at a fast trot down through the pedestrian subway and into the station building; surely there would be something in there. The subway somehow led us directly into a Burger King. There was no way I would eat a Burger King breakfast in Nizhny Novgorod. This was the first time I had ever entered a Burger King and I wasn't going to compound the mistake by eating their food. So we turned around to go back the way we had come.

But! A large bulging crewcut man who looked like he had just returned from

Spetsnatz military duty in Syria was blocking our way, triumphantly pointing at the sign above the doorway: *Vkhod Zapreshchen*! No Entry! He pointed to another door at the other end of the place; there was sunlight through it, and what looked like a street. Would that door take us out of the station? We had no tickets or passports and therefore wouldn't be able to get back in. Our prison guard wasn't going to budge and in a panicky moment I imagined the provodnitsa pulling up the steps and the train departing and leaving us destitute in Nizhny Novgorod. I unleashed a torrent of fast desperate English on our man which seemed to make him think. Unbelievably, he signalled for us to wait, and disappeared around the corner, I think to find someone who could translate for him. Or maybe someone with handcuffs and manacles. Well! We shot through the *Vkhod Zapreshchen* door faster than you could say Nizhny Novgorod and hustled back the way we had come through the subway. Walking very fast, not running, trying to stay below the radar. Wincingly waiting for a bullet in the back, or at least a shout. Out onto the covered platform, suddenly feeling very conspicuous because it was almost deserted as everyone had boarded the train. I had never been so pleased to see our provodnitsa and her steps. If necessary, surely she would be able to fight off the brute. Up the steps, through the coal smoke, and into our cabin. Heads below window level. The Rossiya lazily glided out about seven seconds later. No breakfast today then.

The next three hours passed quickly with us packing to leave the train at Vladimir, a couple of hundred kilometres short of Moscow.

We said *do svidaniya* to our provodnitsa, who cracked a friendly grimace back, and stepped down onto the Vladimir platform right on time at 10.44 am. Into the arms of our next Monkey Business driver whose name I have forgotten, so it probably wasn't Sasha. Vladimir is an ancient provincial town, but we weren't here to stay. Farewell to the Rossiya; we didn't even see it leave. We loaded our luggage and selves into our new man's car and he drove us the 36 kilometres north through rolling hills and farmland to the small town of Suzdal, our own little piece of Russia for the next two nights.

Suzdal announced itself with a profusion of onion domes in front of us, and suddenly we were at the Sokol Hotel. The Sokol was an old tsarist-era bank building with its entrance right on the corner, three storeys, stone and plaster painted cream, with pure white trim. The woman at reception said we couldn't check in until 2 pm, so we left the packs and went out to explore, and find some lunch, and celebrate being on steady ground. We went across the road, through a scrappy grassy empty field, to the long, low nineteenth-century building called Torgoviye Ryady. In English, the Trading Arcades: a few shops selling tourist stuff, vege stalls and a couple of cafés.

We went around the back, walked along the long veranda, and found ourselves at Gostiniy Dvor Restaurant. We sat outside on the veranda in the warm sun with red geraniums among the white columns around us, and a big pewter samovar. There were voluptuous drapes gathered to the columns in case the sun got too bright, which seemed unlikely with angry-looking clouds hanging around. To our left were the five black onion domes of the nearby Friday Church, an imposing white stone cuboid decorated like the sides of a classic wedding cake, with the domes on white towers sprouting from the top. And further to the right, a grassy meadow, adorned with a carpet of white flowers, tumbled down to a corner of the placid Kamenka River which oxbows quietly through Suzdal. And the lunch! Delicious fish soup and meaty pelmenyi dumplings. Afterwards we just sat there marvelling at this paradise we found ourselves in.

Suzdal is grand little backwater 200 kilometres north-east of Moscow. In 1862 the new railway to the east went through Vladimir and bypassed Suzdal completely, thus preserving it from development. By then it was already a time capsule. There are the remains of a fabulous Kremlin that was built, or commenced, in the eleventh century. The town became a major focus point of Orthodox Christianity, with scores of churches and monasteries. Mongol Tartars sacked the place in the thirteenth century, Muscovites in the fifteenth. By the sixteenth century Suzdal was down to 400 households, with more churches than it could ever need. The seventeenth century was bad: attacked by Poles and Lithuanians, laid waste by Crimean Tartars, a devastating fire, and the plague exterminated almost half the population. Since then the population has increased to about 10,000 with 40 amazing Orthodox churches and monasteries remaining. In the eighteenth century, town-planning rules forbade buildings over two storeys, and in the 1960s the Communists earmarked Suzdal as a kind of museum-town, so there was no ugly development. Nothing really happens in Suzdal except for a bit of surprisingly quiet tourism.

Post-prandial sitting-about finished, we wandered back to the Sokol where we could now check in. We were given a lovely little suite right up under the roof with a view across the empty main street to the Resurrection Church and the Kazan Church. Now was the chance for showers for the first time in three days. Suitably cleaned, combed and civilized, we went out again for more Suzdal exploration. We headed for the Kremlin, and its stunning

Russian goat with the Kamenka River and the Elias Church behind, Suzdal.

fantasy, the thirteenth-century Cathedral of the Nativity of the Virgin Mary. It was a tall white-stone and painted-brick edifice topped with improbable deep blue onion domes, encrusted with golden stars and crowned with gold crosses. The interior was a glowing blue gallery of frescoes stretching vertically and astoundingly above us. We carried on down to the Kamenka River. Even though Suzdal is a town or a small city, everything is very rural, with lots of long grass and grazing animals; it doesn't feel built up at all.

There was the suggestion of a boat ride through the town on the Kamenka River, which I thought was a grand idea; we just needed to wait half an hour or so for the next boat. We strolled in the evening light along the Kamenka, taking in fabulous views and reflections of the 1774 red, white and green

Elias Church on the other side of the river. It was all brought to a crashing halt by the weather. Without warning, a sudden gale sprang up. The trees started to wave around crazily and medium-sized branches crashed to the ground. Great clouds of spring seedheads blew everywhere, looking like snow wherever they landed. Forked lightning split the sky, thunder rolled around, and the placid Kamenka erupted into a minor maelstrom. It was actually a little frightening. The promised boat trip was definitely off. We set off back to the Sokol, increasingly quickly as the first large raindrops splatted around us. Fortunately, we had raincoats with us; by the last few metres the rain was solidly pouring down. Up to our sanctuary under the roof, looking out at the suddenly bedraggled town.

We were both very tired now after a lot of walking, and three long 25- and 26-hour days. Rail-lag had finally set in. The storm cleared and we watched the last light of the day fire up the silhouettes and onion domes of the Trading Arcades, the Resurrection Church and the Kazan Church, in a conflagration of orange and purple.

Friday, 31 May

Suzdal

Our first job today was to find the Post Office. We had yak wool to send to Germany. A short way along Ulitsa Lenina, set back from the road by a standard Russian expanse of unkempt long grass, was a long, glassy single-storey building advertising Post Russia, with a Russian tricolore stuck in one of the windows. Anne had no trouble unloading a dodgy-looking tangle of smelly yak wool into a bag and over the counter. I didn't quite believe that it would make it to Germany and then on to New Zealand, but it did, a few months later. Freshly caught up with capitalist ways, the Post Office was also a cornucopia of little bits and pieces to add to your parcel. Anne bought a Lego Post Russia postman for somebody's child; I bought a model Ilyushin-96 jetliner in Post Russia colours for somebody else's child. Or me. In the end I did the right thing and gave it to the child.

Mission accomplished, we walked back across the scruffy green expanse of grass to the main street and carried on to the huge Saviour Monastery of St Euthymius. This is a complex of religious extravagance incorporating several churches, inside a really serious defensive wall. Even the main gate is a church. The original wall dates to the fourteenth century, although I'm sure it has been replaced many times since.

Once we got through the Gate-Church of the Annunciation and into the monastery we were in the large Apothecaries' Garden, where the monks historically grew a wide range of medicinal and culinary herbs. It's still a well-labelled (in Russian) herb garden. We had the place virtually to ourselves and tried numerous nibbles to compensate for our inability with Cyrillic script. The huge thick stone and plaster walls beckoned and we climbed up to the wooden ramparts. Some of the bastions of the wall were tall towers. We ended up in one, where scenes from Soviet movies were being projected on a wall, apparently for our sole benefit. I think the movies were filmed in and around this monastery, hence the connection. There were lots of grand imperial-era uniforms, women in long fur coats, snow, sleigh rides, horses, strutting postures and the like, all accompanied by Soviet cinema music which sounded like imitation Prokofiev waltzes. Fabulously stirring stuff and it detained us for quite a while, until eleven o'clock when the bells of the Cathedral of the Transfiguration of the Saviour were let loose, so we wandered over there for a look.

The Cathedral of the Transfiguration of the Saviour is perhaps the biggest and most impressive of the churches inside the walled compound of the monastery, with a big gold onion dome and several smaller green ones. Inside it were more amazing frescoes and three monks singing Russian liturgical music: a high tenor, a baritone and, being Russia, a real basso profundo. The acoustic of the place was such that just three voices sounded like a whole choir, and they filled the cathedral with their glorious sound. An interaction between the tourists' wallets and singers' hands was expected here, but even at the heart of Suzdal's tourist function there were only about 20 people. Musically replenished, we were back out into the warm sun, passing a little stone chapel which I think held the remains of the eponymous St Euthymius, Sankt Yevfim in Russian.

Any good monastery must have a miserable prison for the incarceration of those who don't toe the line, and here was a grim low white building which served that purpose. For a while in the 1940s it was the place of detention of Friedrich Paulus, the vanquished commander of the German Sixth Army, which the Russians defeated at Stalingrad in 1943. Imprisonment in the Saviour Monastery of St Euthymius was a considerably better deal than most of the one-third of a million men under Paulus' command received.

Back to the wall, we walked around the outside of it above the banks of the Kamenka River, which had completely calmed down again after last night's storm. We stopped for a rest at a bench with a view over the little river and down towards the Convent of the Intercession. Here there was an unkempt, unshaven individual with messy hair and baggy shorts who looked oddly familiar. As soon as he heard me speaking to Anne, he asked me if I was a Kiwi. He, and his wife, who was a little less conspicuous, were from Christchurch in New Zealand. I didn't know him, but now I realized that he looked like any middle-aged Kiwi bloke on a summer holiday, and that was what was familiar about him. I hadn't seen anyone like him in two months. It was an unexpected pleasure to catch up with the two of them for a while, and converse in fluent New Zealand English. They were receptive to classical music, and just a week or so previously had been to a performance of Bizet's

Carmen at the Ulan-Ude Opera which I thought was quite an achievement. Both for them and for Ulan-Ude.

Still, we hadn't come all the way to Suzdal to talk opera with two New Zealanders; I could do that any time at home, so we carried on our way, down to the bank of the Kamenka and across a rickety wooden pedestrian bridge to the white-walled Convent of the Intercession, which we had seen from under the St Euthymius walls while talking to the Kiwis.

Inside the walls was an oasis of greenery and gardens, and the Intercession Cathedral, all white except for black roofs and golden domes, in beautifully restored condition. One or two young nuns were walking about and it seemed a women's world.

The Convent of the Intercession has quite a reputation as a place of exile for Tsars' wives. Basil II's wife Solomoniya, unable to produce a male heir, was banished here in 1525, until her death. In 1575 Ivan the Terrible's fifth wife Anna Vasil'chikova was found unsatisfactory, so she ended up in the Convent of the Intercession until her death. Then in 1698 Peter the Great sent his first wife Yevdokia to the Convent of the Intercession for nineteen years, even though she had produced an heir, Prince Alexei. Peter later became paranoid about Alexei's perceived premature designs on the throne and had him extensively interrogated, which killed the poor boy. Meanwhile Yevdokia, in this convent, had an affair with a Noble army officer named Glebov. Peter was not too pleased when he found out and had Glebov tortured then publicly executed in Red Square in Moscow.

Later we walked a bit further, found ourselves back in the centre of town, if there is such a thing, where a haircut for me seemed a good idea. There was a sign with scissors and a comb on it, pointing into a narrow passageway. We followed the dark passageway, and a glass door led us into an empty but fully set-up hairdressing salon. There was no one here. I tried *Zdravstvuyte!* a few times but got no reply. Through another door to another passageway and stairs, it was looking quite residential but still deserted. A few more timid shouts in bad Kiwi Russian, then I gave up and headed back out towards the street. Almost there, and there was a gruff call and a shuffle behind me. I turned around and was confronted by today's Amazon hairdresser, a tall woman a little younger than me, with several hours' worth of make-up on her face. Bright red lips, heavily coloured eyelids, optimistically renavigated eyebrows. No sign of a smile and the possibly unintentional sneer could have curdled cream. I felt like I'd disturbed Fafnir in his cave. But she was friendly and agreed to cut my hair. Counting on fingers, I showed her that I wanted my usual 4 millimetres on the top, 2 on the back and sides, and away she went with her electric trimmer and scissors, quite slow and careful.

Russian millimetres must be smaller than New Zealand ones, because I ended up looking like a Red Army conscript or, as Anne thought, a contract killer. I was actually quite pleased with the result; it would last a long time. The woman was pleased that I was pleased, and Anne got over her initial shock and was quite pleased too. A hundred roubles. That's NZ$1.90 for something closely approaching a shaven scalp.

Out on Ulitsa Lenina, the sky had clouded over and we got a bit of intermittent rain. Somewhere down towards the Kremlin there was a supposedly excellent restaurant called Chaynaya, close to the Kamenka. We strolled in light rain down the tree-lined Ulitsa Kremlyovskaya, past the rather underutilized horses and carriages for tourists, including a four-wheeled pumpkin for those who wanted to be Cinderella, and eventually found the place hidden out the back of a few closed tourist shops. Chaynaya had a big wooden deck at the rear, seemingly hanging above the Kamenka and river-flat fields and vegetable gardens. The rain was constant now, though fairly light, and we could sit out there under cover. The place was quite casual and comfortable, and almost empty. At the next table was a party of Russians in their thirties: two women and about four or five men. They seemed to have an awful lot to talk about.

So did we as the food came out. It was magnificent in a sort of gastro-pub way. A beautiful tender white rabbit leg steamed in a paper bag, a shashlik of grilled summer veges and an earthenware pot of braised mushrooms and potatoes. The Soviet food queues that I was brought up believing in were a thing of the past. That was all very good, but the real revelation of the evening was the condition of our fellow diners who just got drunker and drunker as the night wore on. There appeared to be some sort of crisis that had to be dealt with. They would divide up into twos or threes and stagger off to a corner to discuss stuff. Then they would come together again and redivide into different groups. The men were physically all over each other, touching and hugging. The conversations became louder and louder and more impassioned, but not violent. The two women kept some control of themselves and didn't move around much; from time to time a man or two would come back to them, as if for advice or guidance. Eventually, the men could hardly stand but still they kept going.

In New Zealand this would be getting threatening and ugly, but somehow here it all felt reasonably safe. Perhaps it helped that we couldn't understand

Torgoviye Ryady, Resurrection Church, Kazan Church, Suzdal.

what they were talking about. At last, it all collapsed into a stupor and, show over, we paid and left. As we walked around the front to the street, two of them were still at it, arm in arm on the front veranda. Now we had seen the fabled Russian drunkenness at work, and it was quite a spectacle to behold. We ambled home through the wet dusk to the Sokol and a thankfully sober night's sleep.

Saturday, 1 June

Suzdal to Moscow

Breakfast on the Trading Arcades veranda for the last time. Cafe Gnezdo Lekarya (Healer's Nest) went to some trouble over its breakfast menu. I had the Hangover Omelette which came with a frightening 50-ml shot of vodka. Now it was farewell to gorgeous Suzdal, and a 40-minute drive south to Vladimir for the twelve o'clock train to Moscow: a Lastochka, a fast commuter train made in Germany, which makes the four-hour trip between Nizhny Novgorod and Moscow, four times a day. Train 729G, travelling 180 kilometres from Vladimir to Moscow in one and three-quarter hours. The only really memorable aspect of the journey was the roughness of the ride. The train might have been built in Germany by Siemens, but the rails were certainly not. We occasionally nudged 155 kph as we bumped and jerked and clattered our way to Moscow, with swaying Russians stoically holding on in the aisles for the whole trip. Arrival at Kurskaya Station in north-east Moscow was the end of eight days spent on and around the Trans-Siberian Railway.

Out onto the street with our final Monkey Business driver, and Moscow slapped me about the head with its vastness. We got onto the Garden Ring, a racetrack which circles central Moscow. The traffic was unbelievable: sometimes recklessly fast, sometimes at a standstill. All the Ladas which had been so completely absent in Irkutsk were here, admittedly struggling to keep up. The Maybach Mercedes in the next lane demonstrated the range of wealth in this place. Occasional areas of high ground revealed a frighteningly huge city spreading in every direction. Twice over the doglegging Moscow River, the road became the twelve-lane Smolenskin Tori. We turned right, into a little laneway by the towering Foreign Ministry skyscraper, and there was our Mercure Arbat Hotel for the next three nights, on the edge of Smolenskaya Square.

We had a comfortable room up about three storeys, in a classic elderly building made modern. The window opened, and you could poke your head out to look into the quiet laneway below, leading to Smolenskaya Square in

Smokers' Market at the Torgovaya Ploshchad, Suzdal…

one direction and the twelve-lane horror of the Ring Road in the other.

By now a late lunch was necessary, and we found a good café out on the

Ring-Road. It was beautifully warm, so we sat outside and marvelled at Moscow. The Ring-Road was so wide it created a real sense of space around us. The buildings were all eight to ten storeys, except for the gigantic 1940s fantasy of the Foreign Ministry Building. Some had higher Stalinist towers on the corners, and all in a massive-brutal sort of style which combined to make you feel very small and insignificant.

The road and its traffic were just extraordinary. The road was a major artery but not a motorway, so there were intersections and traffic lights. I'm not sure if there was a speed limit here, but everybody took off from the green traffic lights like drag racers and got up to about 130 kph before braking for the next red lights. This was super-rich hoon city: Ferraris, Audis, Maseratis, muscle-bikes, all with modified exhausts to make lots of noise.

After we had had enough of the aural battering we got up and slipped around the corner to famous Arbat Street, a fifteenth-century major thoroughfare which has been pedestrianized. We walked a long way; it was quite attractive but full of people and basically a tourist trap with endless shops and stalls selling all sorts of junk, all of it overpriced.

We got to the Moscow Conservatorium of Music, with a bronze statue of Tchaikovsky out the front, and the sound of a clarinettist practising, floating down from an open upstairs window. Inside the foyer, there was a box office and we saw that there would be a concert titled 'The Russian Violin School' on Monday night, so we booked a couple of tickets for that. Over the cashier's shoulder was a large poster for another concert, with a photo of Alexander Lazarev on it, staring at us. Alexander Lazarev was very familiar to us as a conductor of the New Zealand Symphony Orchestra. In the 1990s he almost became our Music Director, his 'price' for the 'favour' was New Zealand citizenship. This was made possible, and all he had to do was supply a birth certificate, but somehow he couldn't do that and the deal fell through. A pity, because I think he was one of the best conductors we ever had.

Anyway, Lazarev's advertised concert was some time off and we would not be here. So we carried on walking; up the road there was a square with a statue of the great Mstislav Rostropovich playing the cello. We were certainly in good company. Rostropovich had played with the NZSO too, in 1988. Anne played with him once, in Rome in the 90s. The statue faced across the square to a small church, which was open.

We went over for a look, and a service was taking place. There were no pews; the congregation just stood around in the corners and against the walls. Music played a big part. There appeared to be no choir, just the congregation singing, but it all sounded pretty professional anyway. It was quite fascinating, and we hung around for a while.

I was rather in need of a sit-down so finding a restaurant seemed a good idea. Georgian restaurants in Moscow filled some fantasy niche in my brain, so Georgian it had to be, and we found a good-looking one in a basement nearby. I was interested to watch Muscovites coming in for a meal after a day at work, couples and bigger groups, meeting and greeting. The food was fabulous. Garlicky creamy chicken, veal shashlik, grilled veges, and a Georgian red wine. Later, we wandered out, and along the tree-lined paths of the Tverskoy Bulvar in the surreal northern twilight, watching Muscovites hanging out in the evening, talking, playing chess and tabletop football, going places. We were going places too: back to the Mercure Arbat to get horizontal.

Sunday, 2 June

Moscow

We met Tatyana, our final Monkey Business guide, for a three-hour walk around Moscow. Tatyana was a friendly, easy-going forty-something woman with a short bob of dark hair. She suggested a tour of the metro, then a look at Red Square. That suited just fine.

A tour of the metro might sound rather odd, but the Moscow Metro is an event in itself, with each underground station seemingly trying to outdo its neighbours with ever more stunning decoration. Forget that it's a mass-transit system with trains in it, the whole thing is a work of art. Surely one of Stalin's better ideas, the first section was opened in 1935. It's now 460 kilometres long, has 276 stations, and carries an average of seven million people every day. It is supposedly the largest underground system outside of China. Many of the stations, and the streets or squares above them, are named after cities or regions of the Soviet Union, and the decoration of the stations reflects that, usually with reference to the Revolution or World War II.

Our local station, beneath Smolenskin Tori, was Smolenskaya, named after Smolensk in the far west. The long gallery of Smolenskaya, platforms on either side, was all white marble columns up to the white plaster concave ceiling. The columns created a narrow long view of marble to the far end of the gallery. Above us were intricate carved rosettes containing detailed paintings of scenes from early twentieth-century peasant life in the Soviet Union. Ornate chandeliers completed the impression of a Palace of the People. On the terminating wall at the far end of the gallery, facing us from a distance, was a life-size mural of a multitude of people in traditional costume, and a couple of military officers, paying adoring attention to a flower-bedecked folk-dancing couple. With a big crowd of real people already in the gallery, the mural looked like a continuation of reality out to the peasants in a field on a glorious sunny day in the Paradise of the Workers, Soldiers and Peasants.

Onto a train then, to another station and another world. Most of the trains were unstreamlined boxes with distinctive strengthening ribs pressed into the

Belorusskaya Metro Station, Moscow.

steel sides. Fast, windy and noisy, they were sort of 1950s utilitarian and a complete contrast to the underground wedding-cake stations. We got off for a look at Kievskaya; here were intricate rosettes containing Red Army heroic scenes from World War II, all done in detailed mosaic. Belorusskaya had ornately plastered pure white ceilings using repeated wheat sheaf motifs; how do they keep it clean? And at Novoslobodskaya one of the illuminated stained-glass panels depicted a pianist, in tails, sitting at a grand piano with the lid up. At Komsomolskaya we entered a world of rich yellowy beige, with ornate coats of arms, hammers and sickles, and just to remind us that Muscovites will never forget, some kind of a winged Soviet angel trampling underfoot a swastika-decorated eagle against a backdrop of Lenin's mausoleum. I think if you used the Moscow Metro on a regular basis you would end up with a thorough grounding in glorious Soviet history.

At Revolution Square we came up from the metro and into Teatralnaya Square. The first thing we saw was a temporary fish market set up in the square for a seafood festival, an odd idea for Moscow which is 650 kilometres from the nearest sea, the Gulf of Finland. Beyond the seafood extravaganza, sitting in self-assured isolation, was the Bolshoi Theatre. Turning around the other way, there was a stone Karl Marx, looking just like that other famous nineteenth-century German, Johannes Brahms. The bird which constantly sat on top of Marx's head gave him a faintly ludicrous appearance. Just over Marx's shoulder, in the middle distance, as if to remind him that the Opiate of the Masses is more durable than his own beliefs, was the red, white and gold extravagance of the Kazan Cathedral, built in 1993 to replace its predecessor, which was destroyed on Stalin's orders in 1936. Around to the left from here, up the road a bit, was the Lubyanka, former headquarters of the KGB. You certainly didn't want to be called there in the middle of the night. If my identification is correct, the place has been given a slightly friendlier paint job since the bad old days.

It was time to grab world history by the horns: through the Voskresenskiye Vorota, the imposing twin-towered gate into Red Square. There was a big police presence at the gate, and baggage x-rays, but it was all quite relaxed. Red Square turned out to be more of a long broad pedestrian thoroughfare than an actual square, along one side of the Kremlin wall, past Lenin's mausoleum, leading the eye and the feet to the mad confection of St Basil's Cathedral, now deconsecrated and a museum. The sight is so outrageous that Disney couldn't have done it better. It demands that you stop and stare at it for a very long time, after which you would still not be able to draw it from memory. Just in front of St Basil's is the memorial to Kuzma Minin and Suzdal's Prince Dmitry Pozharsky, who expelled Polish and Lithuanian invaders from Moscow in 1612. This reminded me that whoever happens to be in the ascendant at a particular time is a transitory state, and the last 400 years of French-German-Polish-Russian history features almost every permutation of that.

We turned back through Red Square to the Alexandrovsky Garden and the Tomb of the Unknown Soldier, all low brown marble with an Eternal Flame, just like the Irkutsk one. But no schoolgirl soldiers, as in the provinces; here we got a small squad of real soldiers who had completely mastered parade-ground techniques. It was Changing the Guard time and these guys, individually and in small groups, were demonstrating their goosestepping abilities.

Even though there were lots of people looking and taking photos, the event had a genuine sombre feel to it. The horizontally resting bronze spear with hammer and sickle in its head, the draped bronze flag and the empty bronze helmet, all lying on the marble plinth next to the flame, had a certain finality and gravitas to them. Tatyana the guide told me that guard duty at the Tomb of the Unknown Soldier is a great honour, but it is reserved for soldiers from the far reaches of Russia. Soldiers from Moscow would be distracted by mum and dad or the cousins coming down for a look at their Ivan on a Sunday afternoon, and that would never do.

Now we needed lunch, and Tatyana had the perfect solution: The gigantic GUM department store. Famous for much of the twentieth century as an exclusive shopping outlet for high-ranking party members, and now open to anyone with enough roubles to swap for Sony, DKNY and Calvin Klein. It's like a vast atrium inside, with mezzanines built around an empty centre and a giant glass roof above. Tatyana took us up to the top, right under the glass, to Stolovaya 57, a Soviet nostalgia café serving the food of 1957 in the style of 1957, to people who mostly were not alive in 1957. They even had a 1957 Soviet-style queue of people stretching a long way out of the door. There was no mucking around, and the queue moved very quickly as stuff was slopped on plates or grabbed out of cabinets. I think they had their own Gulag group of 1957 workers out the back. I joined the roleplay and had Herring in a Fur Coat. That's diced pickled herring covered with grated boiled potatoes and carrots, and mayonnaise, and a red fur coat of grated beetroot on top. Not bad, actually. And a bottle of kvass, a sort of beer made with rye bread.

That was the end of Tatyana; we had had our three hours' worth, and she had done a great job. She told us that there was no point trying to get into St Basil's or the Kremlin, as all tickets would be sold out for today and maybe tomorrow too. That was okay, because we were enjoying looking from the outside, and we had tentative plans for tomorrow — and definite plans for tonight.

Tonight was Bolshoi night. More particularly, Adolphe Adam's ballet

Giselle at the Bolshoi. I think that anybody visiting Moscow has to go to the Bolshoi Ballet. I was initially a little disappointed, as the composer Adam is a bit of a lightweight, and French, not Russian. But in fact *Giselle* is an absolutely classic ballerinas-in-white-tutus-dancing-in-blue-moonlight ballet and this was a chance to see the Bolshoi Company at its best. Bookings were snapped up before they disappeared, and we'd made a commitment for a night out in Moscow, long before we left Wellington. I actually had a suit stashed away in the bottom of my pack for the past two months; it was looking okay now after hanging in the steamy bathroom for a while.

It was time for dinner, and I had had my mind on the fish market in the Square all day. Grilled salmon, new potatoes and capsicums, sitting outside in the warm Moscow evening at a small table on little plastic chairs. This could have been Malacca or Hanoi! Except I was wearing a suit and trying to keep the salmon from jumping upstream onto my dangerously wayward tie. Then across the Square to the Bolshoi, beckoning like a classical Greek temple with its eight white columns.

The auditorium was predictably magnificent, and I'm not sure how it squared with seventy years' worth of Communist Party apparatchiks. The stalls area was all red upholstery on individual hard-backed wooden chairs, and above that six galleries, ornate with gold fronts, red drapes and crystal chandeliers. In the centre at the back was a magnificent Royal Box, two levels high, regally draped and surmounted by a two-headed Russian eagle with a crown on top. A crown. No sign of a hammer or a sickle anywhere. Turning back to the huge stage, the curtain was a riot of red and gold double-headed eagles. The proscenium arch was in a heraldic style with sheet music, a massive lyre, and ten red-flag-draped trumpets radiating out like sun rays. Almost Soviet, but not quite. I can't imagine Stalin ever walking in here.

The ballet was a revelation from beginning to end. The dancers, as expected, were stunning. The risk-taking virtuosity of the soloists and the discipline of the corps-de-ballet was the best you could imagine. It was a beautiful traditionalist production. But it was the orchestra that was the big shock for me because they were just fantastic. They played Adam's simple, tonally undemanding score as if they felt and believed every note, and the result was magic. Real rare magic. We got so much more than what was printed in black on the white pages. Of course, the strings were excellent; they are the Bolshoi. The principal viola stood up in the pit for the famous *Giselle* solo and was exemplary. The woodwind and horns were beautifully expressive. The brass had an easy night, but they were always tasteful and idiomatic and could let fly on the few occasions when they were allowed to. It wasn't the precision that got to me, it was the uncommon level of love and care in the playing. I'm going to remember that performance of *Giselle* for ever.

The Bolshoi was full, so it was just as well we booked. During the interval I had a good look at the audience. I think the majority of them were tourists, and not one of the men was wearing a tie, and hardly any suits even without a tie. A leather jacket or a denim jacket or a jersey seems to be de rigeur for the Bolshoi these days. To think that I dragged all this formal stuff through the tropics and to Mt Everest and back so I could sit with people in leather and denim. It appears the world has passed me by, while I wasn't watching.

Monday, 3 June

Moscow

The morning found us at Novokuznetskaya, just south of the Moscow River, in a bike-hire shop. Three cool young dudes, very friendly, had us sorted with good city bikes in no time and we pedalled off towards the Moscow River just a couple of blocks away, then headed west on the Ovchinnikovskaya Embankment. With everything flat beside the river, the riding was very easy, but we were quite slow, taking Moscow in, on a broad pedestrian boulevard with people in ones and twos and the odd in-line skater. To our left handsome classical six- or eight-storey apartment buildings, and over the river to our right a less attractive cityscape.

Further on we came to the preposterous Peter the Great Monument, a 100-metre-high monstrosity of bronze and steel: a stylized super-tall sailing ship with an outsize Peter the Great standing on it. It was built in the 1990s, supposedly to commemorate 300 years of the Russian Navy, but possibly repurposed from a design for a Christopher Columbus statue that no country in the Americas wanted. We passed the New Tretyakov Gallery, but there wasn't time to go in. Here we had much more expansive views of the river and the city; the enormous gold-domed Cathedral of Christ the Saviour to the north, and far away in the south-west, beckoning above everything else, the 240-metre tower of the Moscow State University, one of the so-called Seven Sisters, seven skyscrapers which Stalin had built in the 1940s and 50s. This one was the tallest building in Europe from 1952 to 1990.

Throughout this part of the day, the Moscow State University was a constant and slowly growing beacon, though we never quite made it there. But we did get to Gorky Park which had been on the Moscow Bucket List: a huge expanse of quite dense greenery interspersed with pathways, gyms, sports fields and outdoor fitness trails.

Over the Pushkinsky Pedestrian Bridge and back again, to the Luzhnetsky Metrobridge, and somewhere around here we turned and retraced our route back to Peter the Great. This time we crossed the narrow Vodootvodny Canal and rode slowly along the Sofiyskaya Embankment, looking directly over the river to the walls of the Kremlin.

This was basically a live postcard of Moscow and simply cannot ever be forgotten. On the left-hand end of the Kremlin wall was the fifteenth-century Vodovzvodnaya Tower, then the huge seventeenth-century Terem Palace, home of the tsars, set behind and above the wall. Further on was the extravagant chaos of gold onions which is the Annunciation Cathedral and the Cathedral of the Archangel, and behind them, filling in the gaps, the Dormition Cathedral. Then, surprisingly close to the river, colourful St Basil's again, just outside the Kremlin. Churches everywhere. With five big cathedrals within a stone's throw of the seat of government, it's hard to believe that the communists managed to suppress Christianity so successfully for 70 years, and its huge revival since 1992 now seems to have been inevitable.

A bit further east, we were lured to the dazzling apparition of the Kotelnicheskaya Embankment Building, another of the Seven Sisters like the Moscow University Building. According to Khrushchev, after the war, Stalin felt Moscow should have skyscrapers like New York, so he ordered the construction of eight. One of them never made it, hence the Seven Sisters. They were really built to impress. Apparently poorly designed, over engineered, and very expensive to build despite the use of Gulag slave labour, they sucked up half the resources which Moscow's postwar housing projects so desperately needed. Still, what an amazing vanity project, and they are fantastic to look at. The Kotelnicheskaya Embankment Building is right on the junction of the Yauza River and the Moscow River. It's 176 metres high, with only 22 usable storeys, but it has quite a horizontal bulk and its bright white stone gives it colossal impact. Near the top are huge stone sculptures of Socialist Man and his obedient wife, surmounted by a tall spire with a star on the top. It was designed as exclusive housing for the elite, but quickly repurposed as multi-family communal apartments. It must have been a great address to have, and I am sure it is solely for the new elite now.

Having exhausted ourselves, it was time to return the bikes and then we walked back over the river to the Kremlin. We checked out the large modern Kremlin tourist booking-hall to see if there was anything available for tomorrow, but the ladies behind the armoured glass were quite unhelpful and rude. No wonder they needed armoured glass.

Outside in the Alexandrovsky Garden I suddenly ran out of steam and had to sleep under a tree for a while. When I woke up things were still not quite right, and I wondered if I had a minor dose of flu coming on. Still, we had things yet to do. We had to get back to the Conservatorium for tonight's 'Russian Violin School' concert, for which we had bought tickets two days ago. We found ourselves strolling along another sylvan pedestrian-only Moscow bulvar, this one with big memorial fountains. It was the long Gogolevsky Bulvar, which eventually took us to the Tass news agency, a familiar name on the façade of a horrible concrete-and-glass Soviet bunker. This was the street of the Conservatorium and we found somewhere for a pre-concert dinner nearby.

Sitting on the footpath with a bowl of pelmeni while the Ladas and Lamborghinis inched past in the congestion, looking at Muscovites going about their after-work business, was a real treat. However, despite the warmth I was freezing cold and tired, and I knew something was coming my way.

Still, the prospect of the concert was very exciting, and over the road we went and into the Conservatorium. Past the hewn white granite bust of Tchaikovsky with baskets of fresh flowers at its base, then up the wide steps along the luxuriantly draped and chandeliered corridors and into the auditorium. You could tell the people here were Moscow people, not tourists. Now we were in the hallowed space of the Moscow Conservatorium Concert Hall. It was almost a religious experience.

This place is famous from all those grainy black-and-white photos on the covers of 1960s Melodiya LP records. I've seen pictures of Barshai, Oistrakh, Richter and Shostakovich in here, so I was very familiar with the wide shallow stage backed by organ pipes, though of course I'd never been here before. Being the pinnacle of Russian musical training, it was sort of the epicentre of Russian music, and there were the great composers' portraits in rosettes above the gallery all the way around the hall. Not only the Germans — Bach, Mozart, Beethoven and others — but Tchaikovsky (of course), Glinka, Mussorgsky, Rimsky-Korsakov and . . . Dargomyzhsky. I had no idea what Alexander Dargomyzhsky looked like, I've never heard any of his music, and I had to use an app on my phone to work out who he was from the Cyrillic.

All this excitement led to the depressing anticlimax of the actual concert. From the first note it was a huge disappointment; the less said about it the better. Anne and I walked out in the interval, which I think I've never done before in my life.

There was a positive aspect to this, though: we now had plenty of time for our final project for the day. Back in Lhasa, Thomas the German from Shenzhen had suggested that while in Moscow we should visit the Ukraina Hotel and go up to the 35th floor to take in the view. With a 35th floor, the

Muscovites

Ukraina is obviously yet another of Stalin's Seven Sisters. It's now known as the Raddison Moscow by all other than old Muscovites.

We took the metro to Kievskaya and followed Google Maps in the dark through a parkland of trails, trees and buildings. There was no sign of a 200-metre-high hotel anywhere in the spooky darkness, until Kutuzovsky Avenue where we walked out of the trees and there was the Ukraina, in its lit-up yellow glory, above us. It is supposedly the tallest hotel building in Europe, and it was massive, glowing in the dark. Stalin was right. You just had to be impressed. Inside, we bumped into a rather well-off looking elderly Greek couple and the lady wondered if we knew the way to the thirty-fifth floor. When I said no, we are from New Zealand and we'd like to go to the thirty-fifth floor too, she fell in love with us even though the old guy wasn't particularly impressed. So she clicked her fingers and a big tough-looking concierge appeared and showed us to the right lift and pressed the special security button for the top floor.

We got out of the lift in complete darkness and straight into a dingy dark bar which may as well have been in the basement, stumbled up a dark stairway to a closed door, and suddenly we were outside in the sky. There were glass safety barriers, but we were outside on the top of the Ukraina Hotel with only the spire above us, and the lights of Moscow at our feet and stretching into the distance. Over to the west was the International Business Centre, Moscow's surprisingly small collection of futuristic-looking modern high-rises. Apart from that pocket, and the Seven Sisters, much of Moscow seems to stay within about ten storeys so the view is not as mind-boggling as Shanghai or Tokyo; it's just a massive spread of light in every direction, with the Seven Sisters popping up as punctuation marks. Very close, just over the river, was the Russian Federation Government House, known as the White House. This is where the coup attempt against Gorbachev in 1991 played out and Boris Yeltsin made his famous speech on top of a tank, thereby opening what seemed to be a path to democracy at the time.

I got quite carried away with photographing the glowing White House in the dark, whereupon a security thug told me I couldn't use my camera, but my iPhone was okay. I chose not to argue. He was much bigger than me and it was a long way down. Back down on the ground we re-entered the maelstrom of the Moscow night. The assault on the ears by endless late-model European muscle cars tearing up their tyres on a Monday night was just extraordinary. We walked away from the surreally shining Ukraina, right past the White House where my camera slipped out of its case again. No need for zoom down here. The enormous Ministry of Foreign Affairs Building, another of the Seven Sisters, right on the corner of Smolenskin Tori and Arbat Street, was our guiding light to the Mercure Arbat. Upstairs to our previously laid neck-level tripwire system and a week's worth of dry laundry.

Tuesday, 4 June

Moscow to Smolensk

Our last day in Moscow. This evening at 6.14, we had a train to catch to Paris. For now, we could go out and see what else we could find in Moscow. Thomas the German from Shenzhen had suggested a trip out to the university, to a particular hill where there is a spectacular view over Moscow, so it was down to the metro again, for a train to Universitet. At Universitet we surfaced on to Lomonosovsky Prospekt and started walking, on the long hot straight road, looking for a hill. It was nowhere to be seen, and now I wonder if we misunderstood Thomas. Perhaps he meant the Moscow State University Tower, yesterday's beacon. Maybe he was a Seven Sisters junkie. Now, suddenly, I crashed. I was in a state of exhaustion and I could not go any further. We dropped into a cool student café to see if a bowl of wholesome Russian soup would pick me up, because we hadn't had breakfast. It gave me the energy to stagger back to Universitet. The train drivers did the rest, all the way home to Smolenskaya with just one change.

At the Mercure Arbat we had already checked out, so I fell into a comfortable chair at reception, surrounded by our luggage. My symptoms were heartburn and exhaustion and a bit of a fever, and I was totally debilitated. There was nothing Anne could do for me, so she went out for a final look around for a couple of hours, then around four o'clock I got to my feet and we began the invalid's journey to Paris, down to the metro for the last time, and a ride to Belorussky Station.

Some years previously, standing on the platform at Berlin Hauptbahnhof one night, I watched a train about to leave for Warsaw. People were sitting in the dimly lit dining car with flowers in the windows, having dinner. I thought, how fabulously romantic: dinner on the train from Berlin to Warsaw — I have to do this sometime. So the train from Moscow, through Warsaw and Berlin to Paris, was a must. Booking it was not easy, especially establishing eligibility for visas for Belarus, which seems to have an oddly complicated relationship with Russia. Once again Monkey Business in Beijing came to the rescue. Although their specialties were the Trans-Mongolian and the Trans-Siberian, they could book the Moscow–Paris Express for us through their British affiliate, The Russia Experience. And RZD, Russian Railways, would give Anne a very substantial 35 per cent discount on her fare because her birthday was within a week of the journey. That almost made up for the exorbitant Belarus visas.

Now we came up from the Moscow metro at the enormous multi-turreted sky-blue Belorussky Station. There was a lot of scaffolding everywhere, and the place was obviously undergoing some kind of refurbishment, so there was no waiting room, and no seats on the platforms. Our train was already at the platform but locked. Boarding would open in one and a half hours, and I was dead on my feet. We found a cramped little café on the platform, bought a drink to establish legitimacy, and I collapsed onto an uncomfortable hard stool. I perched there for an hour and a half, shivering in the warmth, softly moaning to myself. A woman was sitting in the close confines almost on my lap; goodness knows what she thought about this odd-looking gibbering wreck.

After an eternity of discomfort, we were allowed to totter along the platform to our carriage. This train was quite different than the Rossiya. Built in Austria only a couple of years previously, the cabins were light and seemingly roomier. The two beds were already made up and I crashed on one of them immediately. Finally, this was bliss.

I was half asleep as we left Moscow at 6.14. Train 23, travelling 3480 kilometres to Paris in 40 hours: our third-longest train journey, after Irkutsk–Vladimir and Lhasa–Beijing. Farewell to Moscow then, a frighteningly larger-than-life place seemingly designed to make you feel you have the significance of an ant. Although it was culturally more familiar to me than anywhere in Asia is, I found I was always marginally uneasy there. Now that I was feeling sick, I felt I was retreating in defeat like Napoleon. All the way to Paris.

This train was potentially the most comfortable of them all, but something was not right, beneath our feet, and the ride was horribly bumpy. I think maybe we had a square wheel, or at least a badly unbalanced one. I dozed on and off and Anne went to the dining car for a solitary dinner. I just wasn't interested. I awoke again during the long northern twilight and stayed awake for what seemed an eternity of Russian birch forest, which was stunningly beautiful in the evening light. I felt a real sense of contentment cruising through the dense dappled treescape. By now I had developed a nasty hot, sweaty fever and I was dead to the world by the time we reached Smolensk.

Soon after Smolensk we skirted the Katyn forest, near the Dnepr River, where the bodies of 22,000 Polish intelligentsia and social and military elite were disposed of after being massacred on the orders of Stalin. This was in

The Moscow–Paris Express standing at Belorussky Station, Moscow.

1940 after Stalin and Hitler divided Poland between themselves. Despite all the evidence, the Soviet Union consistently denied responsibility, until finally coming clean and admitting the truth, in 1990. A lot has changed since 1990 and the events at Katyn are again an increasing source of tension between Russia and Poland.

The Russia–Belarus border came up at about midnight. It seems there is nothing there. No town, no border post. One minute you're in Russia, next minute you're in Belarus, which used to be the Byelorusski Soviet Socialist Republic, one of the fifteen Soviet Socialist Republics of the USSR. It's rather appropriate that Russia farewelled us with the site of one of its awful atrocities. We had been there for twelve days, longer than in any other country on the Antipodean Express except China, and reminders of its miserable past and doubtful present were always just below, or on, the surface.

Wednesday, 5 June

Orsha to Gotha

Orsha was the first place in Belarus of any size, about 40 kilometres inside the border. During the night I was disturbed by flashing lights that might have been Minsk, the capital, but I didn't really wake up until about seven in the bright morning light at Brest. Its grand sandstone station featured epic carved wheat sheaves and hammers and sickles. The Belarus border police came onto the train, with military jackets, long black boots and miniskirts taken to new extremes. Brest is on the far western border of Belarus, right next to Poland. Here was something I'd forgotten about: the bogies. Our 1.52-metre-wide Russian gauge bogies had to be exchanged for 1.435-metre standard-gauge European ones. The rails in Europe are narrower than the rails in Belarus, Russia and Mongolia. So we were here in Brest for three hours, to reverse the whole process we had gone through at Erenhot on the Chinese–Mongolian border two weeks earlier. As before, the train was broken up and we were shunted into a big shed. At Erenhot I had found the whole process quite fascinating, but this time I went back to sleep and slept through all of it.

Freshly equipped with smooth new European bogies, we departed Brest, and Belarus, crossing the River Bug into Poland and the EU. The clocks went back an hour as we left nine weeks of totalitarianism, and the Cyrillic alphabet, behind. A few minutes later the first stop in the allegedly free world was the Polish border town of Terespol for EU immigration, at the new time of 9 am. Two jolly border policemen with fluent English came into our cabin and we apparently made their day as we handed over our New Zealand passports. 'I haven't seen one of these before,' said one. 'Aha! Exotic documentation!' exclaimed the other. Two thumps of stamps in passports, and we were on our way to Warsaw. I had broken my fever during

Moving Life: Poland with Teapot.

the night and I felt much better, though still very tired, and I slept most of the morning, so I missed out on a lot of Poland. However, I noticed that the ride quality had smoothed out markedly; it was really velvety, making this by far the best sleeper train of the Antipodean Express. Presumably, it was the result of the new European bogies.

We got to Warszawa Wschodnia, Warsaw East, around midday and stopped for a bite to eat. There was a place offering Pavlova which was rather a surprise. I was brought up to believe that Pavlova was a New Zealand invention. The Australians, meanwhile, think they have a claim to it. Now I suppose the Poles think it's theirs . . . A little walk outside was a shock: it was 29 degrees out there! Back onto the train, over the Vistula River, and a cruise through south Warsaw,

all pretty dull and accurately reflecting a fairly ghastly twentieth-century history. We made a brief stop at Poznań late in the afternoon.

Poznań is of some interest to me. It has fluctuated between German and Polish possession for 250 years, since the 1770s when the whole of Poland was carved up between Prussia, Austria and Russia and ceased to exist as a state. Seventy kilometres to the north of Poznań lies the town of Chodzież, population twenty thousand. Chodzież has also been German and Polish over the years. When it was German it was called Chodziesen, or Kolmar. One of my great-grandfathers, my mother's father's father Ralph Michaelis, was born in Chodziesen sometime around 1852. In his teens he moved to England, married the Englishwoman who became my great-grandmother, and they emigrated to New Zealand in 1874. Ralph Michaelis was the earliest of my known relatives, so I feel a tug of interest in the Poznań-Chodzież area. My brother and sister have been in Chodzież looking for records and documents to substantiate Ralph's past. The area's history is awfully turbulent and the trail of Ralph, a German Jew living in what is now Poland, has gone stone cold. I can only be grateful to him for moving to New Zealand.

After Ralph left, Chodziesen schizophrenically traced the Polish tragedy. In 1879 the town became Kolmar in Posen, in the German Empire, Posen being the German name for Poznań. In 1919 it reverted to its old name of Chodzież, in the new Second Republic of Poland. In 1939 it became part of the German Reich, with the name Kolmar, and most of the Poles were cleaned out. In 1945 it became Chodzież again in the Polish People's Republic, and all the Germans were cleaned out. In 1947 the population was down to 7000. So far it is still Chodzież, now in the Third Polish Republic. So much for the sad history of Poland.

It was time to move on from Poznań, and a while later, as we got closer to Germany, we decided to check out the Polish dining car for dinner. We were still eating when the train rolled over the Oder River and suddenly we were in Germany, just like that. We pulled into untidy-looking Frankfurt an der Oder Station and waited there for a bit. It wasn't a very attractive welcome to Germany. We usually arrive in Germany at Frankfurt am Main; this was certainly the wrong Frankfurt. We sat watching nothing much happen in a goods yard, finishing our dinner. On our way again, 80 kilometres to Berlin.

Approaching Berlin in the early evening light through an area of garden allotments, pretty much the first Berliners we saw were a naked couple frolicking in their allotment as we cruised above them on a bridge. There wasn't anywhere for them to go, and it must have left a hundred or so weary travellers asking, 'Did we actually see that?' Anne thought it was a particularly fitting welcome to Berlin.

We stopped in Berlin at Lichtenberg in the east for a while, then moved on. Beside the Spree River, with a glimpse of the former East Berlin TV tower in the sunset, and underground to the Berlin Hauptbahnhof. Somewhere around here I fell asleep, and I wasn't aware of anything else for the rest of the day. I know that we passed through Erfurt at 11.30 pm. I'm sorry I missed it; wonderful French horns used to be made there.

Just before midnight we went through Gotha. Remember the Duke of Edinburgh, Prince Alfred, whom I came across in Hahndorf near Adelaide two months ago, and who also had a Wellington connection? 'O you Prince of German Blood' went the song. Well, Alfred was a Prince of the House of Saxe-Coburg-Gotha. That German blood came from right here. In Gotha. I

hadn't been expecting to run into Alfred again. Saxe-Coburg-Gotha was the family name of the British royal family, from 1901 until 1917 when George V changed it to something a little more patriotic: Windsor.

Thursday, 6 June

Eisenach to Paris

Right after midnight we rolled through Eisenach without stopping. That's a pity; we missed out on Bach, Tannhäuser, Luther, Goethe, the Wartburg and a few seminal moments in the political history of the Germans. Not that we would have found out much about any of that at midnight. Eisenach has to wait for another visit, and we moved on from this curious little hub of historic Germanism, to Frankfurt-Süd. Then south to Mannheim, west over the Rhine and through Ludwigshafen, within two and a half kilometres of Anne's family home and sleeping mother around 4 am. Further on through Neustadt and the hilly Pfälzerwald. This was all picturesque and familiar country, and it's a shame it was dark and I was asleep.

After the original Frankenstein and Saarbrücken we were over the Saar River into France at Forbach and across Lorraine and Champagne to Paris. The Moscow–Paris Express pulled into Paris Gare de l'Est an hour and 20 minutes late at 11 am. The Irkutsk of the west. I think it was our first late arrival since Hanoi. This was our last overnight train for the Antipodean Express and it was a relief to get out and stand on solid ground.

We have a good Kiwi friend in Paris. Mark owns and operates the fabulous Mundolingua, a museum of languages. We weren't planning on staying with him, but he promised to ease us into Paris. As it happened, he was too busy to meet us when we arrived at Gare de l'Est, but he sent along an intern working at the museum, a young Frenchman imaginatively named Pierre.

Pierre found us on the platform and led us down to the Métro. The Paris Métro! First time in my life. So far Alsace, next to the German border, had been my only experience of France, and now in Paris all my senses were working overtime. Onto the Métro, one change at Les Halles, and we came up into Paris at Place Monge, a pleasant enough little square of trees and gravel, with one of those wonderful Paris outdoor markets. A brief walk along narrow streets of tightly packed buildings with handsome windows and wrought-iron work took us to the next oasis, this one circular, basically a large roundabout: Place de la Contrescarpe, with cafés spilling out onto the footpath all the way around.

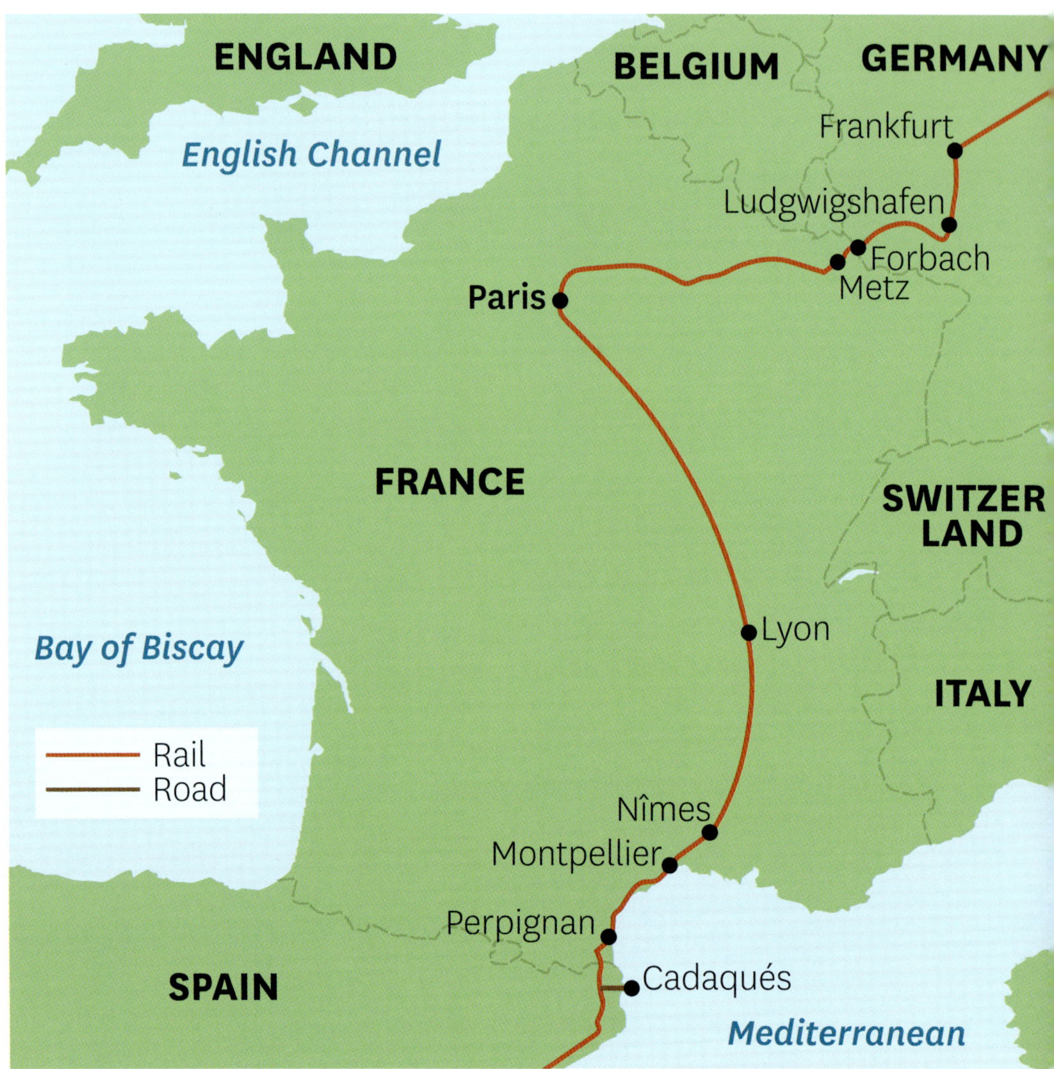

Just to the north of Place de la Contrescarpe, down Rue de Cardinal Lemoine, was our Hôtel des Grandes Écoles, two eighteenth-century three-storey buildings with a beautiful tree-shaded stone-paved space between them, creating an island of repose and quiet in the midst of the city, in the Latin quarter in the fifth arrondissement. There was even the sound of someone practising the piano, wafting down from above, to complete the spell.

We hadn't booked this place; the booking was made by one of Anne's friends in Germany. Anne's four German girlfriends from her student days were all coming to Paris tomorrow to stay in Les Grandes Écoles with us for a reunion and celebration of Anne's birthday. At reception, while filling out all the forms to satisfy French bureaucracy, Mark turned up. A big lunch together seemed a good idea, and the street outside had a plethora of restaurants with enticing

Café de Paris.

blackboard menus, and soon I was enjoying a delicious dish of boudin noir with apple. Meanwhile we caught up with Mark's life and plans. He used to be in the Paris tourist business, so he had plenty of advice to offer. Lunch over, he hotfooted back to work in his museum in the sixth arrondissement near the Luxembourg Gardens, and we went back to Les Grandes Écoles for a sleep.

A sleep on our first day in Paris? Well, we had a big night ahead of us: just as I thought the Bolshoi was compulsory in Moscow, tonight it was the turn of the Paris Opera. After all, I had brought that suit nicely crumpled in my pack with me; it needed another outing to justify its inclusion. We had tickets for Verdi's *La forza del destino*. The Force of Destiny. In Russia we got French ballet, in France we were getting Italian opera. The Paris Opera performs in L'Opéra Bastille, the largest opera stage in Europe, opened next to the site of the infamous prison in 1989. A pet project of François Mitterrand, of *Rainbow Warrior* fame, its seating capacity of 2700 was supposed to bring about a mass popularisation of opera. Like any large-scale artistic endeavour, it has been constantly mired in controversy and Mitterrand, who championed it, ended up saying he didn't like the place. Other presidents haven't liked it either; supposedly neither Messrs Chirac, Hollande, Sarkozy or Macron have ever been inside L'Opéra Bastille, at least at the time of writing. Apparently, it's not just in New Zealand that politicians choose to ignore arts and culture.

But it was good enough for us, and we came up from the Bastille Métro Station and there was L'Opéra Bastille facing us across La Place de la Bastille, all shining curved metal, looking like a large segment of Wellington's Sky Stadium, which we call the Cake Tin. The place was full, and this time there were a few suits and ties. But very few, and I didn't feel anything like the old men who were wearing them. That's it. No more travelling with a suit in future!

And the opera itself: *The Force of Destiny*. I've played the Overture a hundred times, though I didn't know the rest of the opera at all. The Overture opens very distinctively. The brass proclaim one note in octaves, three times, then repeat them: Daa daa daah! . . . Daa daa daah! Then the nervously uneasy first theme starts low and quickly in the violins. Well! Not in Paris. The curtain opened in silence; there were a couple of bars in the orchestra and then the Marquis of Calatrava immediately started singing 'Good Night Darling' to his daughter Leonora. Oh, no. The producer thought he knew better than the composer and left the Overture out. I hate that sort of thing.

But things moved on. The Marquis and his daughter were actually fabulous singers, and so was everyone else who came on after. As well as the soloists, there was a cast of thousands, with a huge chorus to fill that gigantic stage. Apart from the ridiculous beginning it was an excellent production, and later we finally got the famous Overture, to fill in the gap between Act One and Act Two. The orchestra was very good, no complaints there, but there was never a moment of orchestral magic, which we got so much of in Moscow at the Bolshoi. Still, I'm very glad we went. The champagne in the interval was fabulous too.

After the opera we emerged into the night and crossed La Place de la Bastille, under the 47-metre-high July Column commemorating the 1830 Revolution, with its shining golden Génie de la Liberté flapping his wings on the top. We decided to give the Métro a miss and walk home through the Parisian night. Over the Seine and l'Îsle de Saint-Louis, then a long tramp got us onto Rue de Cardinal Lemoine and home to Les Grandes Écoles.

Friday, 7 June

Paris

It certainly wasn't warm on this Paris summer morning. We slowly retraced last night's route along Rue des Fossés Saint-Bernard, to the Seine, and walked along the embankment next to the river, admiring the little river boats. The cityscape of six-storey buildings was beautiful and seemed quite ordered, as if it was all built at exactly the same time in 1820. Soon the back end of Notre Dame hove into view. Well, this was certainly not 1820, in fact most of Notre Dame, from around 1200, is two or three hundred years older than much of the Kremlin. But Notre Dame fits into the city and it's roughly the same colour as everything else.

Today it was still less than two months since the disastrous Notre Dame fire of April 2019. From a distance the Cathedral looked pretty much as it should, but as we got closer we could see that the spire was missing and the whole thing was clothed in extensive scaffolding.

We carried on along the Seine embankment to the gorgeously ornate and beautiful Pont Alexandre III. Actually there was no French Alexandre III; the bridge is named after the Russian Tsar Alexander III. He was the tsar who, after years of Russian procrastination, initiated the building of the Trans-Siberian Railway. Thank you, Alexander III. I don't know if he ever came to Paris, but if he did, it would have been by train just like us. Here, opposite Pont Alexandre III, a fabulous-looking garden leads to the imposing golden dome of Les Invalides, a place of military pomp and La Gloire. I only know it as the venue of the premiere performance of Berlioz' Requiem in 1837. This was where Anne and I parted ways for most of the afternoon. She had to get back to la Gare de l'Est to meet up with her friends, the Gergirls as I like to think of them. They would be arriving from Frankfurt in half an hour. They all needed time together, and Anne needed to speak German, whereas I needed to see tourist things like the Eiffel Tower, so this was a perfect arrangement. I would catch up with the Gergirls later.

I cut inland to get to the Eiffel Tower in a straight line, avoiding a bend in the river. This became an extraordinary and time-consuming zigzag, with the top of the Tower occasionally rearing above the buildings, nearby but unattainably distant. I finally made it, and, yes, the whole business was undeniably spectacular. Of course, I had to go to the top, but I was completely confounded in that. Today you could only go up to Level Two, about a third of the way up; I think because of the wind. So that would have to do then. The lift was packed full of mostly English tourists, and me, and up we went.

Level Two might have been second-best but at 115 metres high it was good enough, and the view of the vast flat spread of Paris was spectacular. Paris is not that big in the world scheme of things, just over two million people, but it looked impressive from up here, a sea of low-rise buildings. I couldn't take too much of this; even though I was well dressed in semi-Everest clothing, the wind and the cold were just awful. I decided to walk down the steps. Getting myself metaphorically lost in that myriad of struts, braces, beams and bolts was the most fascinating part of the day. Eiffel's mind must have been quite something. The way that everything related to everything else was like a huge complex musical score.

Moving on then. What about L'Arc de Triomphe? Over the Seine on le Pont d'Iéna, named for Napoleon's allegedly glorious victory over the Prussians at Jena in 1806. *La Gloire!* Only 5000 French casualties compared to 26,000 for the Prussians. Off the bridge and onto Avenue d'Iéna, just in case I had forgotten. By now the wind, even down at this altitude, was blowing intermittent gales and there were great clouds of dust and pollen which rather upset me. I battled along Avenue d'Iéna in my own little bubble of Gloire, past la Place de l'Uruguay; whatever happened there?

Then there was l'Arc de Triomphe. I tried to picture a blue Volkswagen Kombi van bimbling around that congested roundabout. Sometime around 1978 my parents took a trip all over western Europe in a blue Kombi van. My mother drove all the way and she was very proud that she negotiated the Arc de Triomphe roundabout at rush hour. I heard about it many times and now here I was visualizing it. Then I had the awful realization that my dear old mother, then, was younger than I was now. With that timely reminder of my mortality, I farewelled l'Arc de Triomphe, and the imagined blue Kombi (Blue, my mother called it), and turned around and headed along les Champs-Elysées where I suddenly realized I was starving. I walked into a horribly overpriced tourist-trap café and had a four o'clock lunch of moules-marinière et frites with a beer. It was terrific and perked me up considerably. On down

Paris from 115 metres high. Sacré Coeur Basilica away in the distance.

les Champs-Elysées, and through the Tuileries Garden.

Eventually, I got to the vicinity of the Louvre and its glass pyramid. It was getting too late in the day to start a visit to the Louvre, so I just kept going to the Seine. Where my phone rang. It was Anne. She was only a couple of hundred metres away, coming out of the Louvre with the Gergirls.

So I turned back and we met up. I know them all reasonably well; two of them have stayed with us in Wellington, and the other two will possibly never venture beyond central Europe. It was lovely to see each of them again and catch up, in a cold draughty cloister on the side of the Louvre building. Unfortunately, the Gergirls had thought they were coming to Paris for a summer holiday and had packed accordingly. So they were freezing, and it was time to get back to Les Grandes Écoles. As soon as we got there I crashed onto the bed and I couldn't get up again. With a stomach still full of mussels and chips, I had no need to eat, so the five Gergirls (Anne was one of them now, too) went out for an evening in Paris while I slept like a log.

Saturday, 8 June

Anne's birthday in Paris

We spent the day being Paris tourists, on the streets and in the Métro. We didn't actually go inside any museums, galleries or exhibits; the Louvre was closed, being a Saturday, so that made things simple. I had no desire to queue up for hours anyway. We were all, the Gergirls and I, quite happy to wander the streets for hours just enjoying the ambience of Paris. That ambience was rather cold today, but everyone was wrapped up in whatever warm gear they had. Numerous crêpes, galettes and coffees were had throughout the day, and at some stage we ended up at Montmartre, with a great view from the Sacré Coeur Basilica, reversing the vista I experienced from the Eiffel Tower yesterday.

The huge crowds gathered here were enough to turn you off Paris's tourist attractions, but we found calm and repose on the wonderful Coulé verte René-Dumont. Coulé verte translates as Green Course and that's what it is. A disused long railway viaduct, starting near Bastille, that has been planted out into a narrow 5-kilometre-long garden. What a brilliant way to make use of some obsolete infrastructure. Up to 10 metres above street level, partially walled in, you would hardly know you were in the midst of a big city most of the time, with a few exceptions. It sometimes passes very close to tall residential buildings, so people on the third floor can look straight out into a garden, and at one stage the course went through a narrow canyon between two buildings which must have been built after the railway. The railway was constructed about 1860 and became a park in 1990. The walk was a recommendation from our friend Mark.

On the subject of Mark, now it was time for our private viewing of his language museum, Mundolingua. So we did actually visit one museum. A quick Métro ride to Saint-Sulpice on the other side of town, grabbing a bottle of champagne on the way, got us to Rue Servandoni in the shadow of the monstrous Église Saint-Sulpice, and Mundolingua on the corner. In we went, Mark found enough chairs and plastic glasses and there ensued

A beret in Paris.

a special little private birthday party in a Paris museum for Anne. While we sipped our Laurent-Perrier out of plastic, Mark introduced us to his language museum: an intimate crooked two-storey space just crammed with everything you never thought of in connection with language. Starting at A with the etymology of Abracadabra and moving on from there. We looked at the relationships of languages and dialects, and an extraordinary collection of language technology including an old and frightening Chinese typewriter and a German Enigma machine. To top it all off, there was a remarkably authentic-looking replica of the Rosetta Stone, made by the British Museum. I can't really do it justice; the place was fascinating.

Time to go, we were all, five Gergirls, Mark and I, going out for dinner. First, a quick call-in on Mark's flat just along the road for him to get a coat, so of course a tour of the flat was necessary for Anne and me. Well, that didn't take long, because it turns out Mark lives in a broom-cupboard. No room for a bed, it's up against a wall and he lets it down with a rope and pulley system. Broom-cupboard maybe, but he's in a beautiful part of Paris.

Moving on then, the seven of us walked to the restaurant that we had chosen for dinner. I can't remember its name or where it was; it served upmarket food in a bistro-ish ambience. The food was excellent and the casual atmosphere comfortable. Somehow, unprompted, the staff picked up on the birthday aspect of it all and at dessert time the lights went down, Anne's dessert came out with candles on it, and everyone in the restaurant sang ''Appy Birthday'.

Later we walked all the way home to Les Grandes Écoles, taking in the wonders of Paris on a Saturday night: crowds of drunk young people carousing on the streets, some of them just vomiting where they stood. We could have stayed home on Wellington's Courtenay Place for the same experience, just without the accent.

Sunday, 9 June

Paris to Figueres

Breakfast on Place de la Contrescarpe. The waitress asked where I was from and when I said New Zealand, her jaw dropped in astonishment and enthusiasm. It was almost embarrassing; she thought I was some kind of Angel from Paradise. She still brought me a lousy omelette though. Apparently, real French chefs don't do Sunday mornings. Sort of full but dissatisfied, we wandered down Rue de Cardinal Lemoine and found a big outdoor seafood stall. Oysters, crabs, crayfish, langoustines and everything else that lives at the bottom of the sea. The oysters in particular looked excellent and I had to try a couple. Big, succulent and salty. Right there on the footpath, and predictably expensive. Certainly, none of the Gergirls wanted to try one. A bit further along the road was a strawberry vendor, so I bought a box of strawberries, thinking everybody would like a few. Well, no! No sensible Gergirl was going to eat unwashed strawberries on the streets of Paris. I decided that our dietary patterns were irreconcilable and soon I was full of strawberries as well as oysters — and that nasty omelette.

Now it was time to say sad farewells to poor Anne's distant dearest friends, most of whom she would not see again for some years. We took the Métro to the enormous Gare de Lyon for our TGV, Train à Grande Vitesse, to Figueres in Spain. Finding the right platform at Gare de Lyon was a nightmare. We hoped we were in the right part of the station, but our train wasn't mentioned anywhere until about ten minutes before departure time when it finally appeared on a display, and we queued to get through the ticket barriers just as if for a 1-kilometre Métro trip. Past the pointy-snouted rear power unit and along the platform beneath tall pink, white and blue double-decker carriages towering above us. We found our carriage and climbed upstairs to the spacious comfort of only three big seats across. With not many other passengers, it was all very quiet. Train TGV 9715, Paris Gare de Lyon to Figueres-Vilafant, 734 kilometres in only five and a half hours. Within a few minutes of sitting down we were off, at about 2.10.

Leaving Paris with birthday flowers. Cardinal Lemoine Métro Station.

Our TGV certainly lived up to the Grande Vitesse in its title, and soon we were gliding south towards Lyon at 300 kph. There was surprisingly little of interest to see; small-scale farms whisked by so quickly you couldn't really take them in. South of Lyon we headed down the Rhône Valley and through Nîmes to Montpellier for a quick stop. Around here we began to see the first views of the Mediterranean, our first sight of the sea since the fragrant harbour of Hong Kong four and a half weeks ago, and near Perpignan the east end of the Pyrenees reared up in front of us. I think we were meant to see other mountain ranges on the way, but they were lost in the haze. You couldn't miss the rapidly approaching Pyrenees though, and soon we were in a long tunnel right through them. Out the other side and everything had changed; we were in Spain. Our destination country! We had almost made it to Alaejos.

But first a little holiday with Salvador Dalí on the Costa Brava. Only about 25 kilometres inside the Spanish border, the train stopped at Figueres-Vilafant, the end of the day's journey for us. The train carried on to Barcelona, but we would be spending the night in Figueres. We took a taxi from the station to our hotel right in the middle of town. I think it was only about 2 kilometres, but at €15 it was possibly the most expensive couple of kilometres of my life. The Hotel Duran was a lovely old five-storey place built in 1855 and caringly preserved.

It was already eight o'clock in the evening, but we went out for a quick look around before dark. Down to La Rambla, the main street, and up Pujada dell Castell to the fantastic Teatre-Museu Dalí which Dalí had designed himself as a museum of his life's work and his mausoleum, in the town of his birth. It is extraordinary from the outside with its row of eggs along the top of the wall, but it was all wrapped in red repair-netting which rather spoilt the impact. Not to worry, though, as it was still open during the day, and we had pre-booked tickets online for tomorrow. Somehow, we found ourselves up at Figueres's roof-top level at a vantage point, and there was a good view across the city's jumbled-up mess of terracotta roofs.

The locals hardly think of this as Spain; this was fiercely independent Catalonia, which tried to declare independence from Spain in October 2017. That didn't succeed, and by now the boiled-over tension had receded to a resentful simmer. Till next time. Figueres itself is no stranger to political strife and war. It was a Republican stronghold during the Civil War and was substantially wrecked by German and Italian bombers practising for World War II.

Figueres is also no stranger to fabulous Catalan food. Catalan cuisine is a remarkable and hallowed subset of Spanish cooking. After all, the legendary restaurant El Bulli, considered by many to be the finest restaurant in the world before it closed in 2011, was just 20 kilometres down the road from here, at Roses on the coast. The Duran restaurant in our hotel was, in a more modest way, a shrine to Catalan cuisine — and it was now dinner time. Actually, it was about 9 pm, but the Spanish like to eat late. So, refreshed by our evening walk, it was back to the Duran and into the restaurant: a light, spacious but stuffy place with archways and grand chandeliers. Only a couple of other tables were occupied, by quiet older groups.

Looking at the menu I realized I had hit the jackpot, and the food when it came did not disappoint. I was thrilled to find fideuada with cuttlefish on the menu, a sort of seafood risotto made with thin pasta instead of rice, paired with a light red wine from, of all places, Paraguay. Definitely, a first for me. Then another personal first: Crêpes Suzette for dessert, expertly flambéed at the table by a conservative middle aged waiter who looked like what bank managers and doctors used to look like. Catalan cuisine had absolutely come up trumps.

Monday, 10 June

Figueres to Cadaqués

We started the day with a visit to the Teatre-Museu Dalí, built on the remains of the old Municipal Theatre which was wrecked in the Civil War. The design is largely Dalí's own, and it opened after five years' work in 1974, and continued to expand after that. Quite apart from the art it contains, the building itself is a work of art. Inside were some truly extraordinary huge spaces making clever use of light from above and the side, illuminating giant surreal paintings and sculptures. It was all rather an assault on the senses, and I couldn't take it all in, there was just so much. Some things clever, some surprising, some shocking. I felt numb eventually. Upstairs on higher levels there was a lot of work by other Catalan artists. Finally, dumbstruck and overawed, we got out onto the pavement to find our equilibrium again, rather like disembarking from a long train ride and feeling unfamiliarly solid ground underfoot.

Now it was time to move on to the bus terminal for the 1.45 bus to Cadaqués, just an hour and a quarter away. Cadaqués is a fishing village on the Mediterranean, in the far north-east corner of Spain. It was a little nothing until Dalí moved in for the later part of his life, with his Russian wife Gala, to the next little bay over the hill, Port Lligat. Subsequently it became a, still quiet, destination for artists and movers and shakers, including Federico García Lorca, Mick Jagger, Pablo Picasso and, of course, Anne and me.

When I worked out that the Wellington–Alaejos rail route would take us down the Costa Brava, from the French border to Barcelona, it occurred to me that we had to make a stop somewhere along here. Right now, it was school holiday time in Germany. What about a mid-week weekend in Cadaqués for us with Anne's sister and her family?

The timing and the destination suited everyone, and now here we were on the bus to Cadaqués for three nights of Costa Brava summer holiday with family. The bus was packed. Four seats abreast, it had none of the comfort of the Thai and Cambodian buses. The first half-hour west from Figueres was a dull trundle through bone-dry flat scrubby country to Empuriabrava on the

coast, a resort town of endless manufactured canals and inland marinas, and on to the charming beachside town of Roses, of El Bulli fame. From Roses we were immediately up into the dry rough mountains of the Costa Brava, the Rough Coast. The road was very windy and narrow with rock outcrops and dismal wind-blown drought-stricken vegetation, climbing steeply from sea level to 286 metres in a fairly short distance. There were occasional long views down to the Mediterranean Sea, and eventually Cadaqués sparkled white below us on its bay.

A steep and equally windy descent soon had us pulling into the Cadaqués Bus Station and there were some familiar faces to meet us: sister Ute and her daughter Anna. Anne and Ute took our packs off to our Airbnb accommodation, while young Anna led me to a bar to meet up with her dad Lars and the other daughter, Ilona, and a Spanish beer. We are all very familiar with each other; we had holidayed together in Germany, Switzerland

At last the Mediterranean. Cadaqués.

and New Zealand. Anna had lived with us in Wellington for a term, going to school there. So her English is excellent. Ilona, a year younger, was perhaps on the verge of an English breakthrough, and Ute's English is fine. That just left Lars and me trapped in our own linguistic straitjackets. Lars's English is not very good and my German is even worse, but we manage to communicate quite well through a succession of grunts, nods and gestures.

Beer finished and acquaintances remade, we walked along the waterfront and up narrow, winding pedestrian-only Carrer des Forn to our wonderful house elevated above the town. After the dark narrow alleys, the house was bright and open, with a lovely view over the fairytale harbour, the town and the mountains. First impressions of Cadaqués were of a place built up but unspoilt, locked in a 1950s time warp. No ugly developments, and probably most of the people were tourists, but there was plenty of room to breathe and move. It still functions as a fishing village; the boats are moored out in the bay and there's a fish-processing factory hidden in a valley next to the highway just out of town. There are lots of white sand beaches on the Costa Brava, but Cadaqués doesn't have any of them. There is a beach but it's not really conducive to lying about in the sand. That, plus the terrible road access, seem to help keep mass package-deal tourism at bay.

Having established ourselves with each other, and our home for the next few days, we all went out to check out our surroundings. The whitewashed town was just beautiful in the afternoon light, with a waterfront promenade curving around the bay. We walked a long way, taking in this stunning town and the sea from every angle.

Later we dropped into Es Baluard restaurant for dinner. I was determined to eat out every night. If I was only going to spend five nights out of my entire life in Catalonia, I wanted to make the most of the legendary cuisine while it lasted. Es Baluard was in a pastel blue house next to the omnipresent sea. The restaurant of course specialized in fish and not everyone was as fond of eating fish as I am. Still, we managed to find something to satisfy each of us, especially me. I had Suquet, the local take on Mediterranean fish stew. I tried to make Suquet once, out of a Catalan recipe book, for friends I wanted to impress. It was a miserable fishy failure. Es Baluard's Suquet, however, was simply fantastic. Both words. Simple. Fantastic. With a bone-dry rosat grown in Cadaqués. I had arrived. Even though we finished eating around 9.30, the place was still mostly empty. We spent a long time walking home in the midsummer dusk. I think everybody, from their own perspective whether German or Kiwi, found this to be a uniquely magical place. I had trouble equating it with everything we had done, thinking of evenings in Indonesia and Malaysia, and the fact that we had travelled overland from there to here and seen and eaten everything in between.

Tuesday, 11 June

Cadaqués and Far de Cala Nans

You just could not ignore the view from the deck of our house. Trees obscured most of the town, apart from nearby ceramic tile roofs, but there was a clear view down to the harbour and its profusion of moored boats.

Beyond the boats, a couple of kilometres away to the south, the eye was drawn to a squat pillbox high on a rocky headland above the sea, with no sign of civilisation anywhere near it. This was Far de Cala Nans, the lighthouse protecting the western headland of the harbour of Cadaqués. Its combination of isolation and relative proximity soon had me planning an expedition to walk there. However, apart from Anne, nobody was interested. The teenage girls only wanted to lounge about on couches looking at their phones. Their parents had come here for a relaxing holiday, not a coastal mountain-goat expedition. Anne and I, however, had been sitting on trains for two and a half months and needed some exercise. There was also the fact that this Mediterranean summer's day was 14 degrees, windy and raining.

So we all accepted our different priorities: the family stayed home, and Anne and I set off for Far de Cala Nans wearing selections from the Everest uniform. Cadaqués was pretty quiet on a dull rainy Tuesday morning. Walking along next to the beach we came to a statue of Salvador Dalí, just to remind us he was here. Born in Figueres, he came to Cadaqués regularly in his childhood and youth, for holidays. He certainly looked a bit incongruous today, cutting a dapper pose on the beach wearing a suit and leaning on a cane, with that extraordinary moustache, all in bronze.

Around to the west side of Cadaqués, the town petered out and so did the road, and we ended up on a rocky paved path, well signposted to Far de Cala Nans. We continued past pebbly beaches and rugged rock formations, looking back at Cadaqués across the water. Far de Cala Nans might have been only 2 kilometres away from our house by line of sight, but the reality of the path following the numerous indentations and headlands of this Costa Brava

Lunch. Cadaqués.

was seemingly magnitudes more. From time to time the path took us up the edges of coastal cliffs, and sometimes we cut inland to high ground. From up there the whole area was rocky and bleak; it was most unwelcoming, with only the very hardiest low vegetation. Almost desert-like, despite the grey sky and the rain. Flashes of brilliant red were provided by bedraggled-looking cactuses which had chosen today to burst into stunning flower, glistening with raindrops.

From one vantage point we had a wonderful view back to Cadaqués crowding around the edge of the bay and up the hill, with its plain, unadorned whitewashed buildings, two, three and four storeys high with clay tile roofs. Individually, they were nothing special, but en masse the effect was beautiful.

We did run into one group of tourist walkers, anonymously shrouded in plastic ponchos and going the other way. The precisely laid stone path eventually got us up to the squat lighthouse. After beckoning us all morning it was rather plain and uninteresting, and it was simply too cold and wet to hang around for long, so after a respectful couple of minutes we beat a retreat, returning the way we had come.

Some long time later we arrived back at the outskirts of Cadaqués, cold, hungry and wet. Anne's priority was to get home and out of her wet trousers. For me the priority was a late lunch, a fishy lunch of the sort that Anne wouldn't necessarily like, so I stopped off at a little hole-in-the-wall eatery across the road from the seawall, while Anne carried on home. I took a seat just inside the open door, facing out to the shining wet pavement, the moored boats and the rest of Cadaqués across the bay.

I was actually a bit uncomfortably cold and wet, but I wasn't going to let that get in the way, because the small menu told me I had entered another shrine to Catalan cuisine. I ordered esqueixada, which is uncooked shredded salt cod with some olives and tomato pieces swimming in a bath of olive oil. It was just superb. Western Mediterranean people have made salt cod into a regional staple ever since Basque fishermen found cod in the north Atlantic, salted it and marketed it, 500 or more years ago. There were never any cod in the Mediterranean and nowadays all the salt cod, or bacallà as it's called in Catalan, comes from far-off Norway and Iceland. It certainly made magnificent eating on the Cadaqués waterfront. Just a plate of esqueixada wasn't enough on its own, after a morning of energetic walking, so I rounded it off with six crispy grilled bony sardines, served with nothing other than a lemon wedge and salt, and a glass of lovely pale rosat.

After this I was exceedingly happy with the way the day had turned out. I dragged my freezing wet body home to spend what was left of the afternoon dozing on the couch while everyone else exercised their German predilection for board games, and sheets of rain came and went, out in the bay.

Wednesday, 12 June

Cadaqués and Cap de Creus

We had a booking for all six of us at the Dalí House, just a couple of kilometres away in the next bay to the north-east, Port Lligat. The plan was to continue, after the visit to the Dalí House, for another seven rough kilometres to Cap de Creus at the extreme north-east corner of Spain, then walk home. So our departure this morning had a sense of adventure about it: the whole family setting off for the unknown in the still-crisp morning light.

The Dalí House was in the corner of Port Lligat bay, set off by cypress trees, a complex whitewashed structure climbing the hill just like a miniature Tibetan monastery. The place was originally a collection of small fishermen's huts which Dalí began to amalgamate into one slightly grand residence in about 1930. He and Gala lived and worked here until Gala died in 1982.

We formed up into a party of nine, six of us plus two elderly French ladies and a guide. The first shock was the stuffed polar bear, standing upright to about 2 metres, festooned with necklaces, threateningly operating as both a lamp standard and a cane stand. A polar bear on the shores of the Mediterranean established the general wackiness of the whole place, reinforced by stuffed geese spreading their wings under the ceiling, a kitschy soft toy dog in its basket, and a huge cherry blossom umbrella. Most of the rooms were surprisingly small, but they had windows exquisitely framing gorgeous views out to the bay and the harbour entrance. This was all a look into the intimacy of the Dalí–Gala lifestyle and as such it was much more satisfying than the huge public display in the Figueres Museum. After a good look at the intricacies of this rabbit warren, including the bedroom with two ostentatious blue and red beds surmounted by some kind of golden imperial eagle, we spent time outside on the terraces and in the elevated garden, making use of the well-placed chairs and benches to drink in the view.

After we had our fill of it, we wandered down to the sandy beach. The bay was full of boats, and there were lots pulled up on the sand too. Even Dalí's boat was there, lovingly restored, similar to the others but painted bright yellow. For all its artistic significance, Port Lligat was refreshingly undeveloped. We had seen pretty much everything there was to see. It was the sort of place we could happily sit down and while away some more time, but now we had to find the path to Cap de Creus and get moving again.

Meanwhile, enthusiasm for walking had waned. Ute decided that she would walk back to Cadaqués, get their car, and drive to Cap de Creus. The girls were thrilled that this development got them off the hook too, so mother and two daughters returned the way we had come, while Lars, Anne and I set out for Cap de Creus, about an hour and a half away. Along the flat rock-paved path at the back of the beach there were not many houses, just two non-threatening hotels and a spread of scrubby Iberian bush and the sparkling blue sea with a deep blue sky: it had turned into a glorious Mediterranean summer's day. The road veered inland and eventually got us to our walking track. Now we were in quite a wild landscape of stone tiers, sometimes landlocked, sometimes looking down to the Mediterranean Sea. The paths were carefully and flatly paved with the local rocks, all very impressive and a pleasure to walk on. There were lots of yesterday's cactuses, some of them huge, and many displaying their gorgeous red flowers. Cap de Creus is punctured with lots of inlets, and the track took us down to several of them, mostly rocky but some with sheltered little beaches.

Eventually, the path began a steep climb up the hillside to the lighthouse on the end of the cape, Far del Cap de Creus. We were only 75 metres above the sea but with the view, and the steep climb, it felt like we were on top of the world. Certainly, from the Spanish point of view, we were at the edge of the world. Cap de Creus is at the very easternmost tip of Spain, and it's almost Mediterranean Spain's northernmost point as well. The French border was just 17 kilometres away across the sea to the north-west from here. We had a stunning 270-degree view of the blue Mediterranean: to the south the Costa Brava stretched as far as the eye could see to Cap de Begur, and to the north the Golfe du Lion disappeared into the haze towards Montpellier and Marseilles.

We had spent a couple of hours walking here rather than driving, so we felt pretty smug about it. Ute and the girls were here too; there was a pleasant indoor-outdoor café next to the lighthouse, and it was lunchtime. Sitting outside under an umbrella with that view. Patatas Bravas on the Costa Brava, and calamar. Along with a beer, from a Cadaqués boutique brewery of which the waiter/owner was very proud: Cooperativa Cervesera Cadaqués, with a Modernist clenched worker's fist logo, recalling the International Brigades of

The Dalí House, Port Lligat.
The yellow boat was Dalí's own.

the Civil War. The beer was terrific: a hoppy pale ale that wouldn't be out of place on the Wellington beer scene.

Nirvana couldn't last for ever, though, and as the afternoon wore on we had to get walking again. So goodbye to the car travellers, and back down the steep path, retracing steps through Port Lligat to Cadaqués. Anne and Lars began to revel in their relative youth and left me for dead as I stumbled home with a stomach full of beer and potatoes. Much later we went out for our last Cadaqués dinner. I finally had Zarzuela, a seafood dish with everything in it. Zarzuela is also a frivolous form of Spanish operetta, with everything in it. I'm not sure which came first. I was finished after that and went to bed while everyone else played Die Siedler von Catan.

Thursday, 13 June

Cadaqués to Barcelona

Time to move on: Alaejos, tantalizingly close now, was calling! So this morning we were out the door, door locked, keys irretrievably inside, at 9.30. We walked through the cool, narrow whitewashed laneways, so far untouched by the sun. Quick farewells, then onto the bus for us and we went our separate ways. Ute and family were driving further down the Costa Brava. We were taking the bus back up over the mountains to Figueres, then a train to Barcelona. We drove into Figueres with three hours to spare before the train. Anne went off for a look around and I sought refuge in a leafy, shady tapas bar near the station, to catch up on some salt cod croquettes and a few other fishy bits.

Anne eventually returned and we climbed onto the bright-orange regional train standing at the platform. Soon we were comfortably rattling, not too fast, down the Costa Brava to Barcelona 120 kilometres away to the south-west; about two hours. The train was virtually empty when we left, but slowly filled up at stops along the way.

Barcelona is a seaside city of 1.6 million people: the same as Auckland, but the similarity stops there. We were going to spend two nights here. It's such a revered city that it didn't seem right just to charge through in a hurry. It may be the place that the word 'overtouristed' was invented for, but still I wanted to know what Montserrat Caballé and Freddie Mercury thought all the fuss was about, back in 1987. Tonight, by another of those extraordinary coincidences, we were going to meet up with another member of the family. Felix is Anne's nephew, her brother's son. It just so happened that he was in Barcelona this week, on holiday with his girlfriend and her parents. I have known Felix since he was seven, and here was a chance to meet his first girlfriend.

We got off the train at the huge Barcelona Sants Station, some distance from the central Barri Gòtic where we had booked an Airbnb. The metro was a convenient option, but I thought a bit of a walk would introduce Barcelona to us. Towing our packs, we walked through a rather unattractive modern cityscape for a while before admitting our mistake and descending into a beckoning metro station.

This got us quickly and painlessly to Liceu Station where we surfaced onto the legendary La Rambla, which wasn't at all packed with tourists like it's meant to be. South for a short distance on tree-lined La Rambla, left at the McDonald's, and then we descended into the narrow chaos of the old Barri Gòtic. Finding our way was hindered by the fact that the street we were looking for seemed to have two names. We were on a thin winding thoroughfare carved through the four-storey buildings, Carrer del Call, looking for the similarly named Carrer de Sant Domènec del Call, or was it Carrer de Salomó ben Adret?

Eventually, there was a dark cleft in the wall of buildings to our left, perhaps wide enough for two pairs of shoulders, and there was Carrer de Sant Domènec del Call. It was an unimaginably narrow stone-paved laneway between stone and concrete buildings leaning in towards each other so the gap at the top was even narrower, letting in a mean slit of light. Just crooked enough to show that the builders didn't really care about straight lines. This was the midst of the mediaeval Jewish quarter. We crept into the dark and finally found the right number on a magnificently ornate wrought-iron security gate, and after a flight of stairs found ourselves in a spacious, modern first-floor apartment. Pleasant it might have been, but we certainly weren't going to spend much time in it, and we were out to enjoy the end of the day in Barcelona in no time.

Back onto Carrer del Call, unencumbered now and free to dawdle and stare. Yes, there were tourists about, but it wasn't too bad. This was all that old European inner-city combination of businesses on the ground floor and residences above. A short stroll got us to the open expanse of Plaça de Sant Jaume, the centre of political power for Barcelona and therefore the whole of Catalonia. On one side of the Plaça was the fourteenth-century City Hall; on the other side the Palau de la Generalitat de Catalunya which has been the seat of Catalonian government since 1400.

The unrest leading to the unsuccessful Declaration of Catalonian Independence, less than two years previously, was still evident hanging from the windows around Plaça de Sant Jaume, with lots of posters and banners which didn't need translating. *Llibertat Presos Polítics!* in black, red and

Barcelona.

white. *Democràcia!* in blue, red and white. Lots of Catalan flags, but no Spanish flags. On into the myriad narrow streets beyond the Plaça, and there were so many opportunities to eat and drink! We sat down somewhere for an urgent cava sangría.

Meanwhile Anne had been in phone contact with nephew Felix, and we arranged to meet him and his people at a restaurant in a basement just up the road from our place. Felix the newly minted adult was still childishly happy, which was a bit infectious, and his girlfriend's family were friendly people. After an evening of tapas and talking, we said our goodbyes, strolled the twenty or so metres to our gate and we were finished with Barcelona for the day.

Friday, 14 June

Barcelona

The priority for a day in Barcelona was Antoni Gaudí's preposterous Basílica de la Sagrada Família. Down to the metro, and after a short ride we emerged from the depths to be immediately confronted by the vertiginous shock of la Sagrada Família. From the metro exit the Basílica was just across the road; the four spires of the Nativity Façade soared rather improbably to over 100 metres, crowned by apparently even higher cranes. It was not just the height that was overwhelming, it was the combination of bulk and height. And the busy detail everywhere. The Nativity Façade itself was a crazy grotto displaying every detail of the Nativity. Behind the four completed spires was a forest of more spires still under construction, all representing different figures from the birth of Christianity. The central Jesus Christ Spire will be the tallest when it's finished, at 172 metres, making the Basílica the tallest church building in the world. The attention to detail on every level is an astounding look into Gaudí's busy Christian mind. The Nativity Façade, representing Christ's birth, faces the rising sun. The Passion Façade on the other side, Christ's death, faces the setting sun. The stupendous height of the Jesus Christ Spire when it is completed will be deliberately a little lower than the top of Barcelona's Montjuïc Hill, because Gaudí believed that his own creation should not surpass God's work.

We slowly walked around Gaudí's creation, to the sunset-facing Passion Façade. This is much newer, but still following Gaudí's original design. Compared to the Nativity Façade it is brutal and stark. Construction of la Sagrada Família started in 1882, Gaudí died in 1926, and it should be finished around 2040.

Interestingly, there's a Kiwi element here. For a long time the Senior Architect and Researcher for the project was a man from Christchurch. This is a small country, so of course I knew his father. He was an ex-All Black who, some time in his seventies, decided to take up the French horn. I tutored him from time to time.

First sight of the Basílica de la Sagrada Família.

We didn't go inside the Basílica, as there were huge queues and crowds outside and I was happier to be awestruck by the exterior than to be frustrated by seething humanity. When it all became too much, an ice cream at a table under a tree, still in view of the spectacle, put things right.

We slowly moved back towards the centre of town, on foot, the view behind us dominated for ages by this extraordinary monument to the glory of somebody's God. I think Gaudí, living in a devoutly Catholic country, only just made it in time with his fantasy, as Catholicism is rapidly losing its grip on Spanish society. So who is paying for this wonderful monstrosity? Amazingly, it is the tourists buying entry tickets. Tens of millions of euros every year. There is no government money in the project at all. Actually, it seems that there is no evidence that a building permit was ever applied for or granted, before construction began in 1882. This omission was finally resolved exactly one week before our visit, with the belated issue of a building permit, at a cost of €36 million. At least la Sagrada Família is allowed to be there now.

We took the metro to the Palau de la Música Catalana, one of the most beautiful concert halls in the world. Opened in 1908, it is only three years older than our own revered Auckland Town Hall. But whereas the Auckland Town Hall is all starchy Edwardian Italian Renaissance Revival, the Palau de la Música Catalana is seductively curvy ornate Catalan Modernism. It's built on a congested corner so you cannot take the whole thing in in one go, but there is endless detail to absorb.

There was a personal aspect here, too. The Palau de la Música used to be the home of the Orquestra Simfònica de Barcelona. In the 1980s and 90s that orchestra's Artistic Director was the exceptional Franz-Paul Decker, who was at the same time the New Zealand Symphony Orchestra's much-loved and much-hated Chief Conductor. An absolute master of the German symphonic tradition, he also brought with him an infectious enthusiasm for Spanish and Catalan music, so I was thrilled to see where he picked this up from. He had passed away a few years previously, but I half expected to bump into him in the foyer of the Palau de la Música. And there was yet another connection: one of my NZSO French horn colleagues, an American import, spent time playing principal horn in the Orquestra Simfònica de Barcelona, in the 1990s, with Franz-Paul Decker. But by the time that horn player moved to Wellington, Decker was long gone.

I had had my fill for now and sat down for a bit of tapas and a glass of wine followed by a late siesta back at our apartment, while Anne went successfully shoe shopping — for espadrilles of course. But this was our only Barcelona day and soon I was up and out again, reunited with Anne in new shoes, exploring in other directions.

The ancient Roman wall demanded attention. What an extraordinary spread those people had, away back then. In Roman times in Barcino, as it was then called, there was supposedly a young virgin called Eulàlia who was somehow exposed naked in the Town Square before a spring snowfall covered her up. Roman officials punished her by putting her in a barrel studded internally with knives and rolled her down the street. Her body is now in the crypt of the Cathedral of the Holy Cross and St Eulàlia, and she is a patron saint of Barcelona. There was a fabulous longish view of the fourteenth-century Cathedral across the Plaça de la Seu and the façade was fascinating to stare at for a while, but by now I had seen enough amazing things for today and it was time for our last Catalan dinner. We found a really good, slightly Asio-Hispano fusion place very close to home. What a good idea a day in Barcelona had been.

Saturday, 15 June

Barcelona to Salamanca

Back on the train for the long and rather roundabout trip to Salamanca, almost 700 kilometres away with a change in Madrid. At Barcelona Sants Station, we ran smack into a chance to be victims of crime for the first and only time on the whole Antipodean Express. Coming up out of the metro, Anne followed behind me onto the escalator and then it abruptly stopped. Suddenly, a nice man was helping Anne to carry her pack up the escalator, but he insisted that she hold onto it too. Meanwhile in the confusion he created, he got his hand into her shoulder bag. Fortunately, Anne had spent her youth on the Frankfurt U-Bahn and knew all the tricks. The bag wasn't just over her shoulder, it was also clipped close to her body, on her belt. The thief couldn't easily access her bag, but somehow he got the zip open and he was in there, until she swatted his hand away.

He and his accomplice immediately gave up and sprinted away up the frozen escalator, empty-handed, while I idiotically shouted 'Thank you!' after them for helping Anne, not realizing that they had tried to rob her. Anne had been magnificent. She was always going on at me about being careful with your stuff on the metro; it kind of fell on deaf ears, but I was with her all the way now. There were no police anywhere, and now we could see that at the bottom of the escalators there was a big emergency stop button, just waiting for another enterprising robber.

Sants is a large place and was absolutely humming with people going other places. Unlike Gare de Lyon in Paris, we had no trouble finding our train. The Spanish rail system operated by RENFE has a fabulous high-speed railway network with crocodile-snouted trains rushing everywhere at 300 kph. Everywhere, that is, except Salamanca which is relegated to a slower second-tier regional railway. So now we were on our sixteen-carriage bullet, AVE 3112, to Madrid, Puerta de Atocha. Five hundred kilometres, two and three-quarter hours — that's not mucking around.

Aragon and Castile slipped by in a comfortable blur. There was some classic dry Spanish country and village scenery on the way, and somewhere a fantastic crumbled castle on a hill above a small crumbled village, all the same yellowy colour as the surrounding country. La Mancha was just a bit further south, but the standard images of Don Quixote, Picasso's perhaps, with its blazing sun, would have fitted here perfectly.

We arrived at Madrid Atocha at two o'clock and here we had to find a Cercanías local train to take us to another Madrid station, Charmartín, a quarter of an hour away. At Charmartín, on the northern edge of Madrid, we had an hour and a half to wait for the Salamanca train. After some lunch, Anne disappeared into another shoe shop, soon to emerge with a big grin and something else on her feet. Now that the Antipodean Express had almost reached its destination, all baggage-related caution had apparently been thrown to the winds.

So, newly fed and shod, we took our places on the final train of the journey from Wellington to Alaejos, aboard ALVIA 4969, the 3.55 from Madrid to Salamanca, taking an hour and a half to cover 175 kilometres. It rattled and jumped along, reminding me of the Lastochka from Vladimir to Moscow, seemingly half a lifetime ago. Soon after leaving Madrid things out the window became more verdant and hilly, as we headed towards the foothills of the 2500-metre-high Sierra de Guadarrama, and through the town of San Lorenzo de El Escorial.

El Escorial is the enormous monastery and palace complex where Philip II lived an anchoritic existence, sitting at his small desk facing a blank wall while he carefully plotted his fantastically unrealistic Armada invasion of England in 1588. 'We are in confident expectation of a Miracle,' he said. The blank wall experience was soon repeated on us as we blasted through two very long tunnels beneath the Sierra de Guadarrama, then out onto the arid high plateau of central Spain near Ávila, arriving at Salamanca in the late afternoon.

There was a sense of elation and triumphant achievement here; the arrival in Salamanca marked the end of the rail journey to our antipodes. Alaejos would be just a quick bus trip tomorrow. This whole area, including Salamanca and Alaejos, is the Meseta plateau of central Spain and it's quite high, at about 750–800 metres. Here, the summer can be fierce, with very little rain. The winter is very cold, with snow settling several times a year.

Salamanca is another of those names which resonates through newer parts of the former British Empire. Sir Arthur Wellesley, the Duke of Wellington, leading a combined British and Portuguese army, smashed a French army here in 1812 during the Napoleonic wars. Napoleon wasn't here, as he was setting himself up for defeat in Russia at the time. The Battle of Salamanca supposedly made the Duke of Wellington's reputation as a brilliant offensive general. So, of course, in Wellington, near Victoria University, you'll find Salamanca Road. In Hobart in Tasmania, where I lived in the 1980s, at the base of 1400-metre Mt Wellington, there's Salamanca Place at the old docks. All this, combined with the old school atlas which showed Wellington and Salamanca juxtaposed, meant that I've always been cognisant of Salamanca even though I had no idea what it was like. Now it was time to find out.

The reasonably unattractive modern station was fortunately not too close to the centre of the old town; we walked along successively narrower streets until we went through a passageway and out into the enormous baroque Plaza Mayor, designed and built as one entity, and finished in 1755. Salamanca is famous for the golden glow of its sandstone buildings at dusk, and this was just starting to happen in the Plaza now, at the end of the day. No whitewashed buildings here. We cruised in wonderment over the Plaza and out the other side, and on the rear of the façade we found our base camp for the final push to Alaejos tomorrow, the Hotel Catalonia Plaza Mayor Salamanca.

We checked into a big comfortable room and got back out to the Plaza Mayor just as fast as we could. There were tables and umbrellas laid out around the edges of the Plaza, and a mass of people congregating in the centre. We sat down and ordered a couple of glasses of cava from the usual middle-aged professional waiter. Now we could survey this fairyland creation of a Plaza at our leisure. It was four storeys high all the way around, the bottom floor being a colonnaded gallery of shops and restaurants, and the other storeys, maybe residential with French doors, shutters and continuous balconies. We sat for ages drinking, listening and watching the crowds assemble for a Saturday night in Salamanca.

There was a big rainbow flag flying up among all the official ones, confirming that Salamanca was no longer the conservative place it once was. Politically the opposite of Figueres, Salamanca was a centre of right-wing Nationalist power during the Civil War and on through the Franco years.

Around nine o'clock we found a restaurant just a block away with tables on the footpath and an exhaustive Castilian menu. That means lots and lots of meat, and none of your Catalan fishy stuff. Castilian cuisine is perhaps most noted for spit-roasted suckling pig, just like the Babi Guling of Bali, but without the spice. I'd been looking forward to eating in Castilla y León for some time. This is, after all, the original home of the acorn-eating Iberico

Salamanca. Saturday night in the Plaza Mayor.

pig, as sold to Felicity and me at the butcher's shop at Shau Kei Wan in Hong Kong. So now my meal was a great rectangle of perfectly crisped crackling skin hiding a rack of baby shoulder underneath, with a simple but richly flavoured gravy, and an afterthought of crispy roasted thin slices of potato. This was very enjoyable for a while, then it became relentlessly too rich and fatty, and eventually I had to admit defeat and stop. It even had me fantasizing about becoming vegetarian. Too much to eat, but sitting outside in the warm dusk was superbly atmospheric and memorable.

By 10.45 it was fully dark and we dragged our bodies back to the Plaza Mayor. The place was a throng of people; there was some kind of techno rock playing very loud, synchronized with a spectacular light show projected on the City Hall façade. Probably not exactly what the architects, Alberto and Manuel Churriguera, uncle and nephew, had in mind back in 1755. But here was a brilliant example of a Plaza: a place for people, still fit for purpose more than a quarter of a millennium after it was built. To my mind that's pretty successful architecture.

Sunday, 16 June

Alaejos

Finally, the climax of the Antipodean Express. Today we would realize our goal of standing on the antipodes of the carpet in our living room in Wellington.

We had tickets for the nine o'clock bus to Valladolid, stopping off in Alaejos, a journey of about three quarters of an hour. It didn't take long to get out of Salamanca on the A-62, the Autovía de Castilla, heading north through sometimes undulating cultivated country.

The first sign pointing to Alaejos had me on the edge of my seat. The bus turned off the Autovía and pulled up at a dusty roadside stop outside the shuttered-up Café Bar La Frontera. There was a rough line of two-storey buildings on one side of the road, and a tall wheat silo on the other. No sign of people. This was it. We had arrived. Ominously, only one other person got off the bus. She was met by an unsmiling man in a Mercedes with tinted windows which pitter-pattered its way down a narrow lane and around a corner. We watched the back of the bus roar off until it disappeared. It would be back at 5.15 this afternoon.

Our exploration of the antipodes of Wellington began just behind the Café Bar La Frontera where we came upon some ancient ruins. This appeared to be an extensive archaeological site, with lots of excavation and paths laid for access. There were worn stone and brick walls up to four or five metres high, and old stone pavements. I thought it must all be Roman but, no, it turned out to be the Castle of Alaejos, completed in 1453. It seems to have played quite a role in the fifteenth-century Wars of Castilian Succession. Queen Juana, the 22-year-old pregnant second wife and first cousin of Henry the Impotent of Castile, was imprisoned in the castle around 1462. She escaped in a basket to the sanctuary of the unborn child's father-to-be, who was not Henry, and obviously not impotent. Not even a cousin.

We moved on into the town along the narrow Calle Fortaleza y Plaza. Paved with newish small concrete slabs, with no footpath, the road could certainly accommodate one car but not two. The buildings on either side were two storeys high, made of small flat bricks, with big, shuttered doors and windows, and beautiful wrought-iron balconies and lamp stands. Overhanging eaves were ornately painted on the undersides. It was almost sterilely clean, and not a tree or shrub to be seen — or a person. This all led out onto the Plaza Mayor. To our left was the Ayuntamiento — the Town Hall — and across the street the tall sixteenth-century brick Iglesia de San Pedro. Closed even on a Sunday. The Plaza Mayor, three-storey brick all around, opened up to us; it was quite expansive for such a tiny town, and absolutely deserted and silent. Alaejos was starting to feel like a ghost town. By now it was after 10 am, but the silence and the still-long shadows cast by the façade of buildings gave the place a real early-morning feel.

I was quite familiar with the layout of Alaejos, relative to Wellington. I knew that the precise antipodes of this Plaza Mayor is a spot in Wellington's Green Belt which I used to walk past on my way to work. It's just 250 metres south-east of Government House, the Governor-General's residence. I used to climb that grassy hill every morning and imagine the people in Alaejos standing upside down just 12,735 kilometres away directly beneath my feet, at the opposite end of the day and in the opposite season.

Now, standing in the Plaza Mayor de Alaejos and feeling the right way up, my silent reverie was broken by a whirring sound, and I looked up and saw a couple of bicycles riding into the Plaza, then a few more, followed by a peloton of Lycra, Kevlar and ticking sprockets. They all pulled up on the other side of the Plaza, outside a café which was suddenly displaying a big banner in red and white: Club Cicloturista de Alaejos, with Valladolid in brackets beneath. There might have been a hundred of them, exclusively men. Suddenly, after the deserted silence of a couple of minutes ago, the noise was almost unbearable. The all-encompassing façade, the paved ground and the total absence of any softening vegetation amplified everything. The impact of a hundred noisy Spaniards and their garish high-vis Lycra on this quietly dull urban scene was shocking. But at least there were some people here now! They stayed for about fifteen minutes, then at some unseen signal, like a flock of migrating birds, they were off on their whirring tyres, out the other end of the Plaza Mayor, and the unearthly silence descended again as if they had never been here.

We wandered over to the Town Hall and were surprised to find an Oficina de Turismo in a side entrance. Even more surprisingly, there was someone in there, on a Sunday morning. She was a pleasant young woman with good

English who was quite pleased to have some business. I explained why we were here, and she rather uninterestedly replied, 'Oh yes, I think some people from New Zealand were here a few years ago.' She told us that the reason the Iglesia de San Pedro was closed was that it was having some remedial work done to stop it falling down. Certainly, it didn't look very sound from close up. But not to worry, there was the even bigger Santa Marìa at the other end of town, just 350 metres away, and she would go and open it up for us soon. So we picked up a few glossy brochures about the town and headed off to the Iglesia de Santa Marìa down more narrow brick-lined laneways.

We broke out into the sun on the triangular Plaza de Santa Marìa, dominated by the massive, severe bulk of the church on the north side. A black dome at one end, and a very tall square bell tower at the other. Built in the sixteenth century, and mostly original, the long façade facing the Plaza was a rather austere, simple sort of Classical and Renaissance combination. One of my new brochures described the architecture as '*de una extraordinaria sobriedad*' and it was right. I've seen Presbyterian churches looking happier than this one. The whole building was brick: flat ones which individually present a very small exterior so there must have been millions of them.

The Turismo girl arrived and unlocked the big wooden door for us. Inside was something that had once been grand and awe-inspiring but was now shabby and worn. The sanctuary and apse were splendid in their standard Spanish golden glory, but everything else was looking tired. The plaster arches and ceiling were a mess of cracks and damp spots. The wooden floor wouldn't have been out of place in a shearing shed. Here was the story of Spain, really. This once-magnificent church was built when Spain was the most powerful country on earth, and surely the richest, with all that stolen gold from Central and South America. Now there is nobody, and no money, to preserve the place properly. How is this little town with a diminishing population going to look after all this? And there is San Pedro just up the road, already being worked on. Two enormous churches for a town of 1300 people.

We climbed up into the organ loft, a messy and neglected area with beautifully crafted worn wooden railings and a grand view down to the nave. Up here, looking sad in a corner, was a rather ancient free-standing organ, maybe four or five metres high, with the pipes encased in an extravagant wooden carved cabinet. There was a plush red stool in front of the seven pedals and the locked-away keyboard, but it didn't look like it was still a working musical instrument.

A dodgy-looking ladder got us up to a steep staircase past big cracks in the tower, and soon we were in the belfry with an uninterrupted view over all of Alaejos — a jumble of terracotta roofs dissected by streets so narrow you couldn't see much pavement. Towards the other end of town, the Iglesia de San Pedro reared up above everything, dwarfing the nearby wheat silo we had seen when we arrived. The town we had travelled exactly halfway around the world to see was finite. Beyond its abrupt boundary, in every direction we could see green, brown and yellow farmland stretching flatly to the parched, empty horizon. We decided we should get down and relieve the Turismo girl of responsibility for us.

Now it was time to find what we had come here for, the antipodes of our house. Not wanting to put my trust in international connectivity solutions, I had screenshots from Google Earth in my phone, with the relevant spot marked. Soon we were out in the fields on foot, heading north-west. Our path led us straight to an embankment below the A-62, the Autovía de Castilla, then stopped. It seemed that our destination was in a field on the other side of the Autovía. This was a bit of an inconvenience; how on earth were we going to get ourselves across a busy four-lane motorway where obviously pedestrians were not welcome?

The solution came in the form of a large concrete drainage pipe emerging from under the embankment, about one and a half metres in diameter. There was a pinpoint of light at the other end; surely that was the other side of the Autovía de Castilla, and the promised land. So in rising anticipation, we stooped our backs slightly and entered the drain under the Autovía. At least it was dead straight and we could see the other end gradually getting bigger. We finally burst out into the hot Castilian day and discovered that we were rather trapped by fences everywhere, and now what we could see bore no relationship to the picture on my phone at all. We had gone too far! We should not be on this side of the A-62. I had misread the shape of the fields and the roads and we should have stayed on the other side, closer to Alaejos. We retreated under the A-62 to the Alaejos side.

Now, right in front of us, was a yellow wheat field with a backdrop of the Iglesia de Santa Marìa and the low roofs of the rest of Alaejos. We had found it! In the midst of this field at latitude 41°18'55" north, and longitude 5°12'56" west, was the antipodes of our home. With a thumping heart, and a quick wade into the calf-high wheat, I was there. We had made it. This was absolutely the furthest from home that it was possible to go. A straight line on the earth's surface, from here in any direction I chose, would get me home in 20,020 kilometres.

The goal.
An elated me standing at 41°18'55"N, 5°12'56"W in a wheat field just outside Alaejos.

I stood in the wheat and looked around me. There was a farmhouse at the edge of the field on the other side. I guess they are my antipodal neighbours. I wondered if they knew that people from the other side of the world were standing in their wheat field. An old man strolled along the gravel track; I got a wave out of him. Did he know what I was doing here? There was a younger man in a blue shirt with a broad-brimmed hat, breaking in a fine big black horse in the next paddock. Did he realize the strange man waving his arms about in the wheat field was an Antipodean? Of course, none of them knew or would have cared; I had achieved an entirely personal and self-manufactured goal, and my elation was total.

From the wheat field, Alaejos, just half a kilometre away, dominated the southerly view. It was not an attractive place from here, despite the two tall church towers. To the north, east and west, it was all farmland with a motorway running through it. All quite arid, a few buildings dotted here and there, and a big blue sky crisscrossed with aeroplane vapour trails. Nothing like green, damp Newtown in Wellington on the other side of the world, with pines on the hill above the profusion of tightly packed wooden villas and cottages, and incessant wind.

Having achieved the goal, all that remained was to carry on and get back to Wellington, but first we had another five hours to spend in Alaejos. Feeling quite special, we wandered back into town, unsure of what to do next. A lot of people had appeared on the streets, all well-dressed and going in the same direction. This was our first clue that the place was indeed populated. Of course! It was Sunday! So we followed them all to Santa Marìa, to a church service starting at 12.45. Not being religious myself, I was nonetheless interested to see the Spanish Catholic Church in action. There was a good crowd, and the building looked much happier, more purposeful and alive with a congregation in it. My musical career has given me a reasonable understanding of the Catholic mass in Latin, but this was all in Spanish and mostly unintelligible to me. That didn't matter; we had stood through Orthodox services in Russian, and the chanting Buddhist monks in Tibet certainly hadn't needed translating.

The only disappointment was the music. There was a choir of old ladies who simply couldn't sing, and a badly played organ, presumably the dusty antique we had seen earlier. The congregation made no attempt to sing the right notes, mumbling in a discordant monotone. I had expected so much more from the rich musical tradition of Spanish Catholicism. Anne, however, pointed out that the ancient ladies in the choir just needed some new blood, and there probably wasn't any interested new blood available. I'm sure she was right, and it is very sad. Church music in rural Spain seems to be in the same parlous state as the churches themselves.

At the end of it all, we walked to the Plaza Mayor to see about a late lunch. When we got there it appeared the entire population of Alaejos had the same idea. The Plaza had woken up! There were cafés with outdoor seating all around the sides, and no tables free, so we moved in on a couple of well-dressed elderly women who were quite happy to share their table with us. One of them had good, if slow, English and we had quite a party. When I told them why we were here, it seemed to have no significance to them at all. Perhaps they thought that the earth was flat anyway, and we were heretics. Lunch in Alaejos was the culmination of a dream. The ladies were great to talk to, and there was wonderful people-watching. Most of the people around the Plaza were middle-aged or older. It was now a hot day, and there was no shade in the Plaza, so all the tables were huddled around the edges in the narrow shade

Alaejos. Lunch in the Plaza Mayor with *nuevas amigas*.

Looking south from the bell tower of the Iglesia de Santa Marìa, Alaejos. San Pedro in the 'distance'.

of the buildings. A few trees would have been a blessing. Then, just as this morning with the bicycles, as if at some unseen signal, everyone got up and walked away. It was over in minutes and all that was left of this presumably weekly ritual was the waiters stacking tables and chairs, then silence.

Now we felt we had exhausted the possibilities of this little place which had been my focus for so long. We wandered back to the now-open Café Bar La Frontera opposite the bus stop, with half an hour to spare, and sat next to the street in the sun with a bottle of agua con gas. Our big red bus, the beginning of Alaejos's lifeline to Wellington, appeared on time and we were back in Salamanca soon after six o'clock.

Sunday night in Salamanca's Plaza Mayor was even bigger and brighter and louder than Saturday night; the mass of family groups sitting at tables having a good time had to be seen to be believed. Those Spanish certainly know how to socialize and to eat. We sat down to a refreshing sangría then, later, a victory dinner on the footpath somewhere, in celebration of the fulfilment of a dream.

Monday, 17 to Friday, 28 June

Salamanca to Wellington

Such was the end of Anne's and my rail journey to the antipodes of our home. Our Antipodean Express. Events in the four years since we stood in that Alaejos wheat field have shown that the whole journey was remarkably well timed.

A month after we visited Hong Kong, the place erupted into street violence in protest against a proposed bill allowing extradition to China. By the end of 2019 terror and violence affected places I had visited; politicians and protesters were being attacked and a few people had died.

Moscow became a conflict zone in July 2019 after opposition candidates for the City Council election were disqualified. Record-sized protests brought brutal retaliations from police. Similar protests for similar reasons blew up all over Russia, and even Ulan-Ude was affected.

Catalonia unravelled in October 2019 after nine independence leaders were convicted of sedition and sent to prison. There was wild violence between police and protesters, with over half a million protesters on the streets of Barcelona.

Kangaroo Island in South Australia suffered a holocaust of death and destruction in December 2019 and January 2020 when unprecedented bushfires laid waste to the western half of the island. Two people died, 87 homes were burnt out, 40,000 koalas incinerated and other endangered species taken to the brink, and tens of thousands of livestock destroyed.

Then Belarus exploded in May 2020 when Alexander Lukashenko retained the presidency after apparent electoral fraud on a massive scale. There were at least four deaths, 50 people missing, 1300 injuries and more than 32,000 arrests.

To top it all off, Covid-19 arrived at the beginning of 2020, followed two years later by the Russian invasion of Ukraine, and nothing has been the same since.

Our departure from Wellington at the end of March 2019 was about the latest it could have been, with some certainty of safety. I cannot imagine when, or if, such a journey will ever be possible again.

After Alaejos we stayed another day in Salamanca; it is a magical town with lots to see, not least the first-century Roman bridge across the river Tormes and the twelfth-century Old Cathedral. We left Spain after ten days on Tuesday, 18 June. This involved a very early morning train from Salamanca back to Madrid, and a flight from Madrid to Frankfurt. I had booked the flight because I thought that by now, with Alaejos achieved, we would be thoroughly 'trained out' and would dread the 24-hour rail trip to Frankfurt. I couldn't have been more wrong. As it happened, this train journey through the dawn from Salamanca to Madrid Charmartín was so magical that I wanted it to never end, and I was really sorry to get onto the Lufthansa plane in Madrid.

We stayed in Germany for seven precious days of family and friends and some impossibly hot weather.

On Monday, 24 June we said our goodbyes to Anne's mother, and Germany, and took the last long-distance train trip of the Antipodean Express journey: Ludwigshafen to London St Pancras, 712 kilometres, seven and three-quarter hours. From Ludwigshafen Hauptbahnhof we caught a busy regional train for the five-minute trip across the Rhine to Mannheim. We had passed through Ludwigshafen two and a half weeks before, on the Moscow–Paris Express, so now we were rejoining the uninterrupted rail journey from Ho Chi Minh City to London Heathrow.

Getting off the regional train at Mannheim Hauptbahnhof, we were presented with railway chaos. Our next train, ICE 202 from Mannheim to Frankfurt Airport, arrived at Mannheim late, and at the last minute it diverted to a different platform, but many of the staff didn't know. Total confusion ensued. Eventually, rather frazzled, we found the right platform, and an equally rather frazzled official directed us to the wrong carriage. We weren't allowed to board that one because the air conditioning was broken down. Suddenly, the train was leaving, and we just had to leap on any old where.

After three months of virtually flawless train travel across the world, it took Deutsche Bahn to bring the whole thing almost crashing down on the final day. Still, once we got going things were okay. This was just a half-hour trip on an ICE, the German iteration of the high-speed train, 75 kilometres to Frankfurt Airport. The trains between Mannheim and Frankfurt are always jam-packed, with people sitting on the floor in the vestibules, and this was no exception. The ordered comfort of the Chinese high-speed trains seemed an age behind us.

At Frankfurt Airport, underground, we got off and waited half an hour for our next train, ICE 16 to Brussels Midi, 312 kilometres in just under three hours. This was another Deutsche Bahn train, but everything went smoothly now and we were very comfortable. Pulling into Brussels was a depressing experience; from the train it is a dull, dreary, poor-looking place. But we were in Belgium; we could add another country to our tally for the Antipodean Express, bringing it to eighteen so far. Our next train would be our last, making a total of 33 trains, not including mass-transit intra-city ones.

In Brussels we waited an hour and a quarter for the Eurostar to London. Actual Brexit was still seven months away, but there was already a full British immigration and customs process to go through here in Brussels. It was more thorough than anything since the Mongolian–Russian border. From there we proceeded onto our final streamlined rocket-train. EST 9141, 320 kilometres to London in just over two hours. Sitting in great comfort, eating a good lunch with a glass of sauvignon blanc, while Belgium and then France flashed by in a flat dull blur. The carriage was full of vacuously chattering British retirees returning from holidays On the Continent, alternately entertaining and infuriating.

I've always thought the channel tunnel is a technological marvel, but here in Europe there is no fuss at all and it's just another railway. I didn't even realize we had reached Calais and then suddenly we were in a tunnel. Oh, this must be it! There are a couple of longer and deeper tunnels elsewhere, but the channel tunnel supposedly contains the longest undersea railway in the world, at 38 kilometres. Twenty minutes later we emerged into hazy England, the clocks went back an hour, and it wasn't long till the train pulled into London St Pancras Station, right next to a long bar with people sipping champagne, just behind a plexiglass screen on the platform.

My dearest, oldest friend Mark — not the Paris one, he's a bit newer — was there to meet us, and we would stay the next two nights with him. We spent the rest of the afternoon sitting in Mark's beautiful garden enjoying the English summer, which was definitely somewhat cooler than tropical Germany, drinking too much and catching up with his family. The next day was spent walking around London, marvelling at hearing English spoken all around us for the first time in over two months.

On Wednesday, 26 June we left London and flew home the quickest way we could. We walked out of Mark's front door, a five-minute walk to Finsbury Park Station, and the Piccadilly Line took us all the way to Heathrow Terminal 2 in about an hour and a half. Onto our black and white Boeing 777 then, for Flight NZ 1, Air New Zealand's daily shuttle from London to Auckland. I couldn't help noticing that we taxied past a piece of aviation history, a Concorde parked at the British Airways terminal. The last one made its final flight sixteen years previously; presumably this one has been parked there for that long. It was beautiful.

Up into the air, heading for Los Angeles, a distance of 8700 kilometres in eleven and a half hours. The clocks went back eight hours and we landed into a brilliant Californian sunset. The plane spends only a couple of hours on the ground at Los Angeles to refuel, but the time is all taken up by extraordinarily stringent US immigration procedures, with excruciatingly long static queues and total discomfort. You need an electronic visa, just for a two-hour transit! Of all the many border crossings we did on the Antipodean Express, Los Angeles Airport would rank worst equal with Đồng Đăng and Pingxiang on the Vietnam–China border.

Back on the plane, for some immediate sleep. Ten and a half thousand kilometres to Auckland, twelve and a half hours. We took off from Los Angeles about 11 pm and we only got a sliver of the next day, Thursday, because we crossed the International Date Line in the middle of the Pacific Ocean, so the clocks went forward this time, by nineteen hours. We landed in Auckland in the dark at 5 am.

On entering New Zealand, you must fill out an immigration form: which countries have you visited while you were away from New Zealand? Oh boy. That took a while. We went outside into the wet wintry air and followed the green line to the domestic terminal for our one-hour flight to Wellington, where we arrived at 7.30 am on Friday, 28 June 2019, 89 days after we had departed.

Wellington in the dawn on a winter Friday. A taxi took us home. I went inside, into the living room, dropped my pack for the last time, and stood staring straight down at the carpet.

Index

R

S

T